PENGUIN  **W9-CAW-383**

THE GREENING OF AFRICA

PAUL HARRISON is a freelance writer and journalist based in London. He holds master's degrees from Cambridge and the London School of Economics. He has travelled widely in Africa, Asia, and Latin America, visiting twenty-eight developing countries. His previous books on the Third World and development problems include *Inside the Third World* and *Inside the Inner City*, both published by Penguin. He contributes frequently to the *Guardian*, *New Society*, *New Scientist*, and *The Times*, and to publications of major UN agencies such as the World Health Organization, the Food and Agriculture Organization, UNICEF, and the International Labor Office. He is also a contributor to *Encyclopaedia Britannica*. Harrison's interest in the Third World began in 1968, when he was lecturing in French at the University of Ife, Nigeria. He is married and has two children.

IIED/EARTHSCAN

Founded in 1971, IIED is an international organization established to focus attention on 'sustainable development', or the connections which exist between economic development, the environment, and human needs, particularly in the Third World.

IIED's policy, field service, and public education activities encompass health and living conditions in Third World cities, forestry and land use, renewable energy, fisheries, and sustainable agriculture. Each issue has global implications, but IIED's approach is attuned toward practical case-by-case analysis and collaboration at the grassroots level. In 1986, IIED worked with national leaders, the private sector, non-governmental and private voluntary agencies, and the people affected by these problems in twenty-five countries worldwide, especially Latin America, Africa and Southeast Asia. IIED runs an environment/development information unit, 'Earthscan'. This produces a wide spectrum of materials available to the public.

IIED is funded by private and corporate foundations, international organizations, governments and concerned individuals.

For further details write to:

EUROPE
3 Endsleigh Street
London WC1H 0DD
England
Tel: 01-388 2117

NORTH AMERICA
1717 Massachusetts Avenue, NW
Washington DC 20036
USA
Tel: (202) 462 0900

LATIN AMERICA
Corrientes 2835, 7° piso
1193 Buenos Aires
Argentina
Tel: (541) 87 2355

PAUL HARRISON

The Greening of Africa

Breaking Through in the Battle for Land and Food

INTERNATIONAL INSTITUTE FOR
ENVIRONMENT AND DEVELOPMENT – EARTHSCAN

Ex Africa semper aliquid novi

PENGUIN BOOKS

PENGUIN BOOKS
Viking Penguin Inc., 40 West 23rd Street,
New York, New York 10010, U.S.A.
Penguin Books Ltd, Harmondsworth,
Middlesex, England
Penguin Books Australia Ltd, Ringwood,
Victoria, Australia
Penguin Books Canada Limited, 2801 John Street,
Markham, Ontario, Canada L3R 1B4
Penguin Books (N.Z.) Ltd, 182–190 Wairau Road,
Auckland 10, New Zealand

First published in Great Britain by Paladin 1987
First published in Penguin Books 1987

Printed in Great Britain
by Collins, Glasgow

Set in Baskerville

For Bim and Sam

Africa is always full of surprises
Pliny the Elder

Contents

Foreword

When Julius Nyerere, President of Tanzania, was first elected to the chair of the Organization of African Unity, he advised the world in his inaugural speech that 'Africa is in a mess'. Two decades later, despite massive infusions of aid and the adoption of countless new 'strategies for development', Africa is in a worse and escalating 'mess'.

The poignancy of hunger and suffering in that magnificent continent is constantly projected through the television screens of those of us who live in the Northern hemisphere. Thousands of learned reports and expert analyses pinpoint widespread ecological collapse across large tracts of territory. Bankers wrestle, unsuccessfully, with Africa's debt crisis. Wars and civil wars reinforce environmental breakdown so as to ensure that almost one African in every one hundred is a refugee. With some 73 coups since 1945, political instability makes capital investment a seemingly risky business. Meanwhile, Africa's population time-bomb is actually exploding.

Africa, of course, has experienced droughts throughout her history. But droughts do not automatically lead to famine. When growth based on exploitation supersedes prudence, good husbandry and the re-cycling of local resources, then communities living at the margins cannot survive. That seems to be one of the inexorable rules of nature, even though human populations – north and south – have still to learn the lesson.

In the South as a whole, the aid machine which seeks to stimulate development – without always defining what is meant by that word – has had mixed success. In Africa ecological crisis and aid failure seem to reinforce each other. The relationship is incestuous, the one interacting with the other. Despite the presence of over 80,000 expatriate aid workers in Africa, we still seem to learn little from our mistakes. We continue to make them decade after decade.

There is, however, another side to this well-rehearsed litany. It is an important side. Any African field worker knows that throughout Africa there are many success stories within the development process, however spectacular some of its failures may be. That many of these successes – perhaps the majority – tend to be small-scale and localized is not surprising. But they *are* successes, some unequivocally so. They shine as veritable beacons of hope.

So early in 1985 I invited the writer Paul Harrison to talk to my colleagues in IIED about how we might identify those successes and expose them before a much wider public. *The Greening of Africa* is the result. I think it is a rather remarkable result.

What, we wanted to know, makes for success in the development of Africa? What are the characteristics of such success? Are they culture-specific, or can they be replicated on an ever-widening scale? Are they limited to certain sectors, but absent from others? Can success attend the efforts of African governments despite their limited resources? If good aid is achieved by people rather than simply for people, what does that say about the current style of development pursued by the multi-lateral donors?

Paul Harrison set off to research the questions, and tentative answers, in the field. Paul knows Africa well. He has researched and written about Third World issues over the years. But here was another opportunity to come to grips with daily reality, at the village level of survival.

The Greening of Africa is an exciting book. Ranging from Zimbabwe's successful family planning programme, to Oxfam's water harvesting project in the Yatenga district of Burkina Faso, it should act as an inspiration to the heroic endeavours of countless Africans and their aid agency partners. While it offers a valuable introduction to Africa's problems and potential, I hope, too, that senior policy makers in governments and the major donor agencies like the EEC, the World Bank and the UN system of specialized agencies will not only read this book, but will argue and debate its findings as they seek to improve their own success rate. Balance is the real issue – not growth alone. How do we secure the latter within the former?

Without the imaginative support of US AID, UNICEF and UNEP, which put up the research grant in the first place, IIED could not have seen this project through. Distribution in Africa was facilitated by a grant from the Lambo Foundation. We are grateful to those agencies, as we are to the many advisors and consultants who gave Paul Harrison their time in Africa and shared their knowledge and experience without reservation.

In many ways Africa is a richly endowed continent. Despite droughts and famines, it still has vast reserves of soil, water and vegetation, a superb flora and fauna, and a mosaic of human populations whose cultures and histories are as vibrant as any in the world. *The Greening of Africa* demonstrates how, in concrete and specific terms, those human populations can live and develop in balance with their natural habitats. At the end of the day that is all Africa – or anyone else – really needs.

Brian W Walker   August 1986
President IIED

Preface and acknowledgements

In June 1985, Brian Walker, President of the International Institute for Environment and Development, asked me if I would write a book on Africa's success stories in development.

With IIED's agreement, I narrowed the scope to what seemed to me the most critical problems: Africa's crisis of food and environment. If we could find projects that succeeded in these spheres – so far most notable for failures – it might be possible to identify ways out of the present crisis.

The 'success stories' were identified with the help of development agencies and experts. In general I required at least two positive assessments before including a project. Altogether I visited sixteen projects and four major international research centres in six countries: Zimbabwe and Kenya (September 1985); Niger, Burkina Faso and Nigeria (November–December 1985); and Ethiopia (January–February 1986). I visited sites and spoke to beneficiaries as well as project leaders. Because only outstanding projects were selected, the balance of comment is of course favourable, but where projects have shortcomings I have not hesitated to mention them. Nor have I always accepted project managers' own views on the degree of success or the reasons for it.

I owe a particular debt to the creative individuals behind the most successful projects, for this book is in large part a distillation of their practical wisdom. I would like to pay particular tribute to President Thomas Sankara of Burkina Faso, as well as Amare Getahun, B. T. Kang, Bernard Lédéa Ouedraogo, Carl Gösta Wenner and Peter Wright.

Funding for the research came from USAID and UNICEF, with special assistance from the World Food Programme in

Ethiopia. None of these organizations had any editorial control over the content; indeed, even IIED gave me complete editorial freedom. The opinions and analysis are my own and not necessarily those of these organizations. The people who helped me on the ground or with documents are too numerous to mention, but I should like to express particular thanks to Didier Allely, Mahmoud Burney, Amare Getahun, Simon Gillett, Marsha McGraw Olive, Oumaru Ouedraogo, Dennis Panther, Kathy Rorison, Stephen Sandford, and Frances Vieta.

The following people reviewed parts of the draft and their comments helped to improve the final text: Michael Ahearn, Nikos Alexandratos, Robert Berg, Steve Dennison, Arne Erikkson, Gerald Foley, Les Fussell, Amare Getahun, Simon Gillett, Thomas Grannell, John Heermans, B. T. Kang, Rattan Lal, Gerald Leach, Steve Long, Jibade Oyekan, Tim Resch, Kathy Rorison, John Rowley, Richard Sandbrook, Lloyd Timberlake, Brian Walker and Carl Gösta Wenner. Drawings are by Cathleen McDougall (Diagram) and maps by Tony Garrett.

IIED provided a congenial and supportive environment and gave me an entirely free hand. Ross Pumphrey was indefatigable in pursuit of US documents, Carol Lambourne fast, efficient and patient in typing and retyping, Richard Sandbrook and Brian Walker encouraging and tolerant in fostering and promoting the project. I should like to thank them for giving me the opportunity of writing this book.

Finally I should like to thank my wife Alvina and my son Sam for tolerating prolonged absences and long hours at a difficult time.

PART ONE
The Challenge

Do not borrow off the earth
for the earth will require its own back
with interest

Swahili proverb

1 The dimensions of crisis

The 1983–5 drought that hit 30 countries in Africa caught the compassion of the West as never before. For months on end television screens were dominated by the infernal images of emergency feeding camps, of burials and wailings, of children with wasted limbs and haunted, gaunt faces.

For Africa itself the immediate crisis was an even more far-reaching affair, an apocalyptic upheaval that left almost no aspect of life untouched.

In Niger the land was stripped bare when the Harmattan wind blew down from the Sahara. For months on end the sun hid behind thick veils of dust. Drifting sand barred roads and ramped against house walls. For the first time in living memory the mighty Niger river that curves majestically through the capital, Niamey, dried to a string of stagnant pools. Some three million people, half of Niger's population, were hit by drought. Fulani and Tuareg pastoralists were pauperized as two thirds of their herds died. By June 1985, some 400,000 nomads had moved into the cities.

In Ethiopia almost eight million people were affected. Farmers in the central highlands sold off one by one first their sheep and goats, then young cattle, mules and asses, then their cows, and finally even their draught oxen at half the normal livestock prices, while the price of scarce grain trebled. In 1982 six 100-kilo sacks of grain could be bought for the price of an ox – enough to feed two people for fifteen months. By 1984 an ox fetched only one sack, enough for one person for five months. Then people sold their ploughs, their hoes, ate their seed for the following year, turned to boiled roots and leaves, and sent off husbands and sons in search of work or help. Along the major highways, from Eritrea and Tigre in the north to Harerghe in the east, they worked on massive conservation projects for food hand-outs. In the badlands of the north, where farms are cut off on narrow plateaux between gouged-out ravines and chasms, people depended on airlifts of emergency food, or abandoned their homes and trekked for the camps.

In the neighbouring Sudan, the lowest Nile flood for 350 years was recorded, and the fertile land within 30 kilometres of the great river dried up. Many farmers ate their seed, jeopardizing planting for the following year: when survival is at stake, today has to take priority over tomorrow. Some ten million people were affected – almost half the population. Among them were no less than 1,400,000 refugees from Ethiopia, Chad and Uganda, who had fled from the frying-pan of civil war into the fire of famine.

Southern and Eastern Africa were not spared. In Zimbabwe the maize harvests in 1982 and 1984 were down by a third on 1981, and in the disastrous year of 1983, down by more than two thirds. In Botswana over half the population was dependent on food aid.

In March 1985, the peak of the crisis, it was estimated that 30 million people were hit by the drought. Ten million of them were forced to abandon their homes in search of food. Up to 24 countries were simultaneously affected, in a vast scythe stretching from Mauritania across the Sahel to Somalia, down East Africa to Mozambique, and back across to the west coast in Angola.

By the end of 1985, good rains had returned to most of Africa except Botswana and Lesotho, but in ten other countries lack of seeds, late planting, or the continued disruption of war threatened to prolong food shortages. Farmers who had eaten their seed grain, or sold their tools and draught animals, faced immense problems in starting up again.

By April 1986 FAO's danger list had halved again, to only six countries: Angola and Mozambique, paralysed by civil wars; Cape Verde, facing its third year of drought and Botswana facing its fifth; Ethiopia, still reeling from the after-effects of drought, disruption and continued civil war; and Sudan, where, despite record harvests over most of the country, several provinces faced continued deficits and the government lacked the cash or vehicles to buy, transport and distribute the surplus to the needy areas.

Over most of the continent, the immediate danger had receded, and with it went much of the concern and sense of urgency that had galvanized people in the North and in Africa. But a year, or even a run of years, of good rains should not lull Africa or the world into a false sense of security. The famines of 1983–5 were not a sudden, isolated disaster, but rather the visible symptoms of a deeper illness. In the same way that stress on the human body can precipitate an acute outbreak of a long-standing disease, so two or three years of bad rains brought to a head Africa's long-term chronic problems. The return of better rainfall allowed the more sensational symptoms to subside. They will certainly erupt again and again until the underlying syndrome is treated.

Africa's crisis is fourfold. First, there is the food crisis: the gradual decline of production of food and of cash crops per person, with no compensating rise in non-agricultural output that could finance food imports. This has been paralleled by the poverty crisis: a gradual and inexorable increase in the numbers of people suffering from absolute poverty and malnutrition.

Africa's exports have fallen in value, while the cost of imports of manufactures and (until 1986) of oil rose. This has led to the debt crisis, a crippling burden, which places in jeopardy all conventional development efforts up to the end of this decade.

Finally, overshadowing all the rest, there is the environmental crisis. Africa's soils and vegetation are being degraded and impoverished at an accelerating rate. If these processes continue unabated, Africa's future will be grimmer even than her recent past.

Unlike the rest of the Third World, most of the continent is not developing, but regressing. Africa is now, and will remain for at least the next two decades, the greatest challenge of world development.

THE HUMAN DIMENSIONS

For almost two decades now, food production in Africa has failed to keep up with the growth of population. Africa's population is

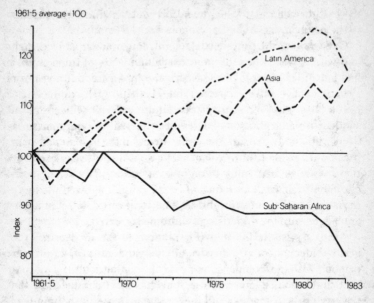

1961-5 average = 100

Per capita food production 1961–83. Source: World Bank

currently growing at the rate of 3.2 per cent a year, faster than any other region, North or South, has ever experienced. The very high birth rate of 47 per 1,000 is 24 per cent higher than Africa's nearest rival, South Asia, and shows no sign of decreasing.

Total food production has, in fact, grown – but not fast enough to keep pace. So food production per person has fallen – by around 12 per cent between 1965 and 1982, just before the recent drought began. In all other developing regions it rose, by between 6 per cent in the Near East and no less than 49 per cent in Asia.

All but five of 38 sub-Saharan countries[1] suffered this decline – indeed there were seven countries where the drop exceeded 20

[1] The total number of countries mentioned for various figures varies, as data is not always available for all countries. When data from North Africa are included, the terms 'all Africa', 'Africa as a whole' etc are used; when North Africa is excluded, the terms 'black Africa' or 'sub-Saharan Africa' are used.

per cent between 1969–71 and 1981–2: Ghana, Botswana, Lesotho, Mauritania, Somalia, Angola and Mozambique. Economic mismanagement, prolonged drought, and warfare played a part in these cases – but only in intensifying a general trend. Even in the most favoured regions of Africa, food production per person fell by 8 per cent between 1969–71 and 1981–2. While Latin America and Asia have become almost self-sufficient in cereals, Africa has grown more and more dependent on imports and food aid. Back in 1963 she grew no less than 96 per cent of her cereal requirements. 1985, the year the drought broke, saw Africa's biggest cereal harvest ever, yet it met only 78 per cent of her much-expanded needs.

Cash crops are often blamed for Africa's food crisis, but there is no Africa-wide evidence of them doing well at the expense of food crops; indeed, the overall acreage of cash crops fell, and yields and per capita production of most cash crops fell even more steeply than food.

Thus farmers, who make up seven out of ten black Africans, saw their cash incomes fall at the same time as food production per person declined. Poverty and malnutrition are the main growth sectors in Africa. A recent International Labour Office survey found that in Asia and Latin America the proportion of the population who could not meet their basic needs, for a minimal diet, shelter, clothing and so on, fell between 1974 and 1982. Only in Africa did the *proportion* rise, from 68 per cent to 69 per cent. Because of population growth the increase in *total numbers* was quite steep, from 205 millions in 1974 to 258 millions just eight years later. Average per capita incomes in sub-Saharan Africa have been stagnant or falling; between 1980 and 1985 they dropped back by almost 2 per cent a year, under the combined onslaught of world recession, cuts in government spending and drought.

Malnutrition is the most acute physical expression of absolute poverty: it is poverty imprinted on human flesh and bone. It drags down people's productivity, lowers their defences against disease, and stunts growing bodies and brains. Until 1982, thanks largely to aid and loans, Africa was able to delay the inevitable nutritional consequences of declining incomes and declining food production per capita. There was a very slow improvement in average daily calorie intakes, from 2120 calories per day in 1961–3, to 2260 in 1979–81 – still 4 per cent less than the level needed to maintain good health and economic activity. Calorie intake then began to fall, to 2230 calories in 1981–3 and certainly much further in the subsequent drought. In 1980–2 there were still 29 sub-Saharan countries out of 44 where average calorie intakes were below requirements – and 16 where intakes were actually lower than they had been ten years earlier.

The armies of the hungry have also been increasing. The numbers not getting the average of 2360 calories a day needed for good health and work are probably the same as the numbers of absolutely poor – around 260 millions, or seven people out of ten. But millions of Africans are not even eating the minimum of 1600–1700 calories a day required just to maintain the human body. These people are, quite literally, starving. They numbered 81 million in 1969–71 according to FAO estimates. Ten years later – even before the drought began – their ranks had swelled to 99 million. Malnutrition and disease work together in fatal combination. The body, weakened by undernourishment, is more exposed to illness – and vice-versa. One African child in seven dies before its fifth birthday. UNICEF estimates that some 3.75 million under-fives died in Africa in 1983. In 1984, with drought intensifying malnutrition, the death toll rose to 4.75 million.

Poverty and malnutrition in Africa are overwhelmingly rural phenomena. Urban incomes in industry and services (mainly government employment) are four to eight times higher than incomes in agriculture. In Tanzanian cities, only one person in five lives in absolute poverty; in the countryside, two out of three do. In Zambia, 24 per cent of town dwellers are absolutely poor – against 52 per cent of rural families.

In Asia, and especially in Latin America, rural poverty is often due to landlessness and the unequal distribution of land. A few African countries, such as Kenya and Zimbabwe, have a very unequal land distribution, surviving from the days when whites seized vast tracts of tribal areas. But in most of Africa, the distribution of land is much more equal. Only 7 per cent of agricultural households were landless in 1980 – well under half the Asian and Latin American levels.

Africa is, overwhelmingly, a continent of smallholders. Two thirds of all holdings are below 2 hectares, and nearly 96 per cent cover less than 10 hectares. Thus poverty and malnutrition in Africa are primarily afflictions of small farmers, and even more so of their wives and children. They are due to the inability to produce enough to meet family needs for food or cash, to the expropriation of a part of the value of production to subsidize cheap food and jobs for city dwellers, and to the low prices of Africa's cash crops on the world market.

THE LONGER SHADOWS

The food crisis and the poverty crisis affect the immediate present. The financial and environmental crises portend an even gloomier future.

The roots of the financial crisis reach back to colonial days when the Europeans exploited Africa as a source of cheap raw materials. That is largely what Africa still remains. Her industries grew very rapidly in the early years of independence – industrial output expanded at the heady rate of 15 per cent a year in sub-Saharan Africa between 1965 and 1973, faster even than in Asia. But then, over the following decade, industrial growth slowed to less than a tenth of that rate – representing an actual drop in *per capita* production of 1.6 per cent each year. By 1982, industry's share of gross domestic product was only 17 per cent, half the Asian level. Manufactures made up only 7–8 per cent of non-fuel exports – hardly any advance over their share in 1965.

So Africa is still heavily dependent on exports of primary commodities – minerals and cash crops – largely in an unprocessed state. In 1982, 87 per cent of the merchandise exports of low-income African countries were primary commodities. Only 13 per cent were manufactures, a quarter of the average for all low-income developing countries.

The world economy has not favoured Africa. The real price of most of her commodity exports has fallen for more than two decades – and the decline has been accelerating. For 18 major commodities exported by Africa, the price in constant dollars fell by an average of almost 4 per cent a year in the ten years to 1983. World recession was one factor. Another was the rise in protectionism in the West, especially the Common Market, where subsidies encouraged farmers to overproduce. Competing agricultural products from abroad faced import barriers, and surpluses were dumped on the world market. Between 1973 and 1983 products like sugar and vegetable oils, which are also grown in Western countries, saw their prices drop three times faster than products like coffee, tea and cocoa, which grow only in developing countries.

Meanwhile the prices of Africa's imports of manufactures and oil were rising, so her terms of trade deteriorated. In 27 out of 32 African countries surveyed by the World Bank, a given quantity of exports bought fewer imports in 1983 than in 1970.

Commodity prices also grew more unstable over the 1972–83 period. For example, coffee prices fluctuated each year by an average of 36 per cent above or below the trend – four and a half times more than in the 1960–72 period. The brief commodity booms of the mid-1970s brought windfall gains in foreign exchange and government revenue to many African countries. Most governments spent the proceeds rapidly. When the inevitable busts followed, governments shied away from cutting back and frustrating the expectations they had roused. Instead, they borrowed heavily. Sub-Saharan Africa's debts soared from $17.8 billion in 1976 to $43.6 billion in 1980 – a growth rate of 25 per cent a year.

Governments hoped to repay later, during the next bull market. Then, in quick succession, the second massive oil price rise in 1979 raised their import bills; world recession cut their export markets – between 1980 and 1984 Africa's exports fell by 7.4 per cent a year in volume; and drought cut agricultural production and increased the need for food imports. Debt repayments from current earnings became more and more difficult. New loans were taken out, not to finance investment, but simply to repay old debts, and the total owings continued to climb, reaching $80 billion in 1984.

Now that famine has, at least for the moment, retreated, the debt crisis is the most serious immediate problem that Africa faces. Arrears have mounted, debts have had to be rescheduled over a longer period, and lenders are reluctant to commit themselves any deeper. Meanwhile the cost of servicing the debt has been growing even faster than the debt itself: between 1978 and 1983 the cost of interest and capital repayment almost trebled, reaching $8 billions a year in 1984. This burden of debt service has eaten up a growing share of Africa's shaky export earnings, from only 5 per cent back in 1970 to 33 per cent in 1985. The World Bank expects debt service to average $10 billion a year over the 1984–9 period. Debt service will consume a massive 26 per cent of export earnings in the last half of the present decade.

Aid has not plugged the gap. It has stayed more or less stagnant since 1980 in cash terms, hovering around $7.2 billion. In real terms, it has declined. Between 1982 and 1984, total net flows of resources to sub-Saharan Africa plummeted by two thirds, precisely at the moment of greatest need, from $8.2 billion to $2.75 billion. Private financial markets, which supplied a net $4.2 billion in 1982, were by 1984 actually taking out a net $480 million. The World Bank estimates that aid and debt relief over the 1986–90 period must average 30 per cent above the 1984 level if Africa's progress is not to be jeopardized seriously. On current plans and projections, there will be a gap of no less than $2.5 billion a year between Africa's foreign exchange requirements and her likely receipts.

Africa has mortgaged her future financially for the next four or five years. Imports have had to be cut, government spending programmes reined back, development projects cancelled or curtailed. In Ghana, health expenditure per person in 1982 and 1983 was one tenth of its 1974 value. Niger's investment programme, running at $60 million a year before 1980, was cut to $2 million thereafter.

The financial crisis is the most severe Africa has ever faced, but it is dwarfed by the continent's deepening environmental crisis. Overall, some 3.7 million hectares of forests and woodland are disappearing every year, according to FAO figures. In West

Africa, 4 per cent of the closed forest is being cleared each year. More serious still is the gradual removal of trees scattered among farms and pastures, crucial in protecting productive land against erosion. Not only trees, but the whole vegetation cover of the soil is gradually thinning out. As populations grow, more and more bush is cleared for agriculture and expanding herds of livestock weaken plant cover.

Once the vegetation cover is removed, Africa's fragile soils are exposed to her winds and her battering rains, and erosion is inevitable. Losses of 20 to 50 tonnes of topsoil per year are not uncommon in cultivated areas. Everywhere soil is being lost far faster than new soil is formed. In many areas degradation takes dramatic forms: the creation of deep gullies, of crusts that water cannot penetrate, of rock-hard layers of laterite that hand-tools or plant roots cannot pierce, of shifting sand dunes that swamp villages and fields. According to UN Environment Programme estimates, a total of 742 million hectares – more than a quarter of the whole continent – is in the process of becoming useless for cultivation, undergoing severe or moderate desertification.

Deforestation and soil erosion are undermining the very resources on which African farmers and their families depend. Crop yields are falling, fuelwood is growing scarcer. And the processes are self-fuelling: land degradation means even weaker plant cover, leading to accelerated degradation.

Africa's environmental crisis will deepen and perpetuate her food, poverty and financial crises. It threatens not just the hope of progress, but even the hope of survival.

2 The harshest habitat

It has become fashionable to pin the blame for Africa's predicament entirely on man-made factors, from an unjust international economic system to the oppression of women, from cash crops to the urban bias of government policy, from colonialism to corruption and overpopulation.

Reality is more complicated still. All these factors play larger or smaller parts in a much wider drama in which traditional societies and technologies interact with modern interventions, on a stage set by the African environment and the world economic order.

It is possible to look at the African tragedy from two points of view: first, there is the logic of Africa's basic human ecology, the way in which the continent's distinctive climate and soils interact with the traditional agricultural system under conditions of rapidly growing populations. This logic itself is potentially destructive. Africa's environment is so sensitive that when a certain level of population density is reached traditional methods begin to degrade the soil on which all future production depends. We shall examine this process in this chapter.

Second, there is the way in which the modern sector of governments and aid agencies have tried to break into that basic logic. As the next chapter shows, they have largely failed to halt its fatal progression – and in many ways have intensified it. This has come about partly by neglect, partly by exploitation, and partly by misguided attempts to introduce Western approaches that were quite unsuited to Africa's very different conditions.

The logic of traditional agriculture and the ineptness of modern interventions combine in a chemistry which is dangerously corrosive of Africa's uniquely sensitive environment.

African is different. Her climate, her soils, her geology, her patterns of disease, all pose problems of a severity that most of Asia and Latin America do not have to face.

A glance at any globe shows Africa's peculiar geographical predicament. The Equator neatly bisects the continent, which is squarely exposed to the glare of the sun like no other region. In the tropics the sun's rays operate at full power. The year-round high temperatures and high humidity in the rainy season provide ideal conditions for pests and diseases – of humans, of crops, and of livestock. These not only affect the quality of life, but depress the productivity of labour, of land, and of animals.

One disease in particular, trypanosomiasis, has had a profound effect on the whole pattern of agriculture in Africa. This parasitic infection, often fatal to cattle, is carried by the tsetse fly. Most of the humid and sub-humid zones are affected, an area totalling around 10 million square kilometres, or about half the non-desert area of Africa as a whole.

Agriculture in Europe, and most of Asia, is based on mixed arable and livestock farming. The cattle provide draught power for ploughing and load-bearing, protein from dairy produce and meat, and fertilizer in the form of manure. Fields that need a rest from cereal cropping can be used to grow fodder. In Africa south of the Sahara, livestock are fully integrated into crop farming only in the highlands of East Africa, which are too cool for tsetse, and a few semi-arid regions, which are too dry.

In the humid zone most livestock except for dwarf sheep and goats are virtually absent. In the sub-humid zone and parts of the semi-arid zone, livestock are at risk in the rainy season. Farmers who possess cattle often entrust them to nomadic pastoralists, who drive them into areas free of the disease – so here, too, full integration is impossible. The separation of livestock and arable farming acts as a barrier to intensification of agriculture in Africa. Since so much soil cultivation and load-bearing is by human hand and head, the separation is a major cause of another distinctive characteristic of African agriculture: the shortage of labour.

THE WAYWARD RAINS

Africa's location has given her a problematic climate. Her wettest areas are sited around the equator (map, p. 346). Here the sun evaporates vast quantities of moisture from sea and land, and as the warm, vapour-laden air rises into cooler reaches of the atmosphere, it condenses into clouds. The latent heat of condensation is released, driving the clouds even higher and producing the towering, four-mile-high nimbus that glower over Africa's rainy season. As the rising air cools, it sheds most of its moisture as rain.

Then the updraught levels out, spreading away to the north and south. It sinks down again, roughly around the tropics of Cancer and Capricorn, and as it descends, it heats up by compression so any remaining moisture is less likely to condense out. In these regions we find the broad belts of aridity that span Africa – in the North, the vast Sahara; in the south, with its much narrower span between two oceans, the smaller Kalahari and its extension through Botswana and southern Mozambique.

As the earth's orbit tilts first her northern hemisphere, then her southern, towards the sun, the rainy zone shifts northward and southward, bringing seasonal rains. They last eight or nine months closer to the equator, grading down to a mere two or three months further away. Some areas have two rainy seasons as the wet belt passes over them twice.

This seasonal seesaw creates the characteristic bands of vegetation that cross Africa north of the equator (map, p. 347). First, the humid forest zone that gets 1500mm or more of rain each year – a level endured in Europe only in her uplands. Rainforest is the natural vegetation of the humid tropics; when cleared for agriculture, tree crops like cocoa, oil palm and bananas, and root crops such as cassava and yam, are cultivated. The humid zone accounts for some 14 per cent of Africa's land area.

To the north is the sub-humid zone of open woodland and savannah with rainfall between 600 and 1200mm. Here the rainy season lasts between four and eight months. This is the major area of cereal production, with maize in the moister areas, and sorghum in the drier. Sub-humid areas take up 31 per cent of the continent.

Moving still further north, we come to the Sahel, dominated by grassland with sparser trees and shrubs. Rainfall here is below 600mm, and the season lasts a mere three to four months. Millet, a tall, spear-like cereal that shoots up and matures in only eleven weeks, is the dominant crop. Semi-arid areas take up some 8 per cent of Africa's land area.

The largest expanse, accounting for almost half the continent, is of arid and desert zones. Here the rains are too short even for millet, but a seasonal flush of grasses on the desert margins allows pastoralists to eke out a precarious nomadic subsistence.

In Southern and Eastern Africa, the closer proximity of the oceans, and the more hilly terrain, complicate the picture. Here the major rainfall zones are not arranged in neat parallel belts, but vary with height. The mountainous areas enjoy the highest rainfall, the lowlands the least. In Kenya, for example, driving north from Nairobi to flamingo-fringed Lake Nakuru, you pass in a matter of kilometres from the lush green farms of the Kikuyu highlands down to the tawny grassed plains of the Rift Valley floor, pastured by wandering Maasai herders.

Africa's rains do not fall gently and evenly. They come predominantly in convective storms. Clouds form rapidly, shed their loads, and disperse. Rain that falls in torrents is less useful and more destructive than the gentler rain of temperate zones: more is lost in run-off, less filters into the ground.

Around the equator, skies are overcast for much of the year and photosynthesis is less efficient. Further to the north and south, dark grey storm fronts give way suddenly to clear blue skies and wilting, withering sun. Much of the moisture is lost through evaporation and transpiration of plants. The annual cycle is like the same shift writ large. The dry season that follows the wet is a period of heavy stress on plants. All annual and shallow-rooted plants die out completely and seeds do not begin to grow again until the rains come. The first downpours hammer down on bare soils, unprotected by vegetation, so run-off and erosion are greater. Then there is a sudden explosion of life. Weeds rocket up, organic matter preserved over the dry months begins to break down, and there is a brief flush of nitrogen in the soil.

These factors mean that the dates when farmers plough and plant are crucial: if they are even a few days late, the weeds will

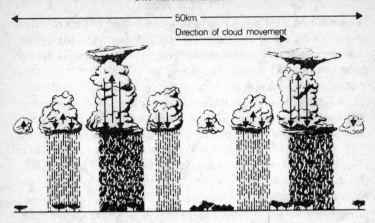

Rainclouds form rapidly and shed their loads in sudden downpours, creating a spotty pattern that scatters drought or plenty in villages only fifteen kilometres apart

get a head start on the crops and outcompete them, and the free fertilizer of the nitrogen flush will be lost. Everything has to be done all at once, and on farms where there are no draught oxen, or where the adult males are working away from home, planting is often delayed, and yields are lower.

Perhaps the most decisive feature of Africa's rainfall is its unpredictability. Even in weather stations only a few kilometres apart, where the *average* rainfall is similar, the rainfall patterns for any given month may be entirely different. One village may be drenched, its neighbour remain bone dry. I remember a dramatic skyscape in the western highlands of Kenya, just before the short autumn rains – to the east, a wall of grey nimbus reaching to the ground, yet right in the middle, a light blue doorway of clear sky, bone dry; to the west, a horizon of fleecy white cumulus on blue, yet with three separate nimbus clouds shedding their load on widely scattered villages.

Dry episodes can come along at any time in the farming calendar. If they coincide with crucial times, when seedlings are establishing themselves or grain is maturing, they can ruin a crop even in a year where total rainfall is quite adequate. At the extreme, one area may suffer drought while another, close by, is

well watered. In 1985, decent rains returned to most of Ethiopia
– yet the region of Harerghe suffered its third year of drought.
The province of Bale had good rains – yet one village I visited,
Alesedestu, was surrounded by a fifteen-kilometre-wide patch of
continued drought.

In the humid regions, rainfall in any given year may be 15–20
per cent up or down on the norm (map, p. 348). As you move
into drier areas, the variation increases right up to 30–40 per
cent in the Sahelian zone and higher in the drier pastoral areas.
The shifts from one year to the next may be even more dramatic,
ranging on occasions from 100 per cent above to 90 per cent
below the average.

Even areas with relatively good rainfall, between 1000mm and
1500mm a year, can expect spells of two successive years of poor
rainfall between two and four times in fifty years. By an unjust
twist of geography, the less rain an area gets, the greater is the
variability of rainfall: the risk of drought is added to the rigours
of aridity. Areas averaging 400–600mm of rain can expect six
droughts of two years or more in every half century. Below
225mm, they may get eight or more. A recent FAO survey found
that two thirds of the continent's land area faced high or very
high risks of drought spells – four or more every fifty years.

The Sahel faces even greater dangers. The recent drought
lasted around seventeen years – the longest and deepest on
reliable record. But prolonged droughts are no novelty to the
area; there have been others, in the early half of the nineteenth
century, in the first two decades of the twentieth, and in the
1940s. US climate expert Sharon Nicholson has shown that
between 1906 and 1980 there were three prolonged episodes of
rainfall above or below the median, lasting eight to fifteen years
each. In the Kalahari desert of Southern Africa no such episode
lasted longer than five years.

Clearly special factors are at work in the Sahel. A number of
mechanisms of the global weather machine have been advanced
as candidates: an intensification of the convection activity around
the equator; the slowing down of the easterly jet streams that
blow across the Sahel, shedding turbulent eddies in which
most storms form; or far-flung 'teleconnections' linking ocean

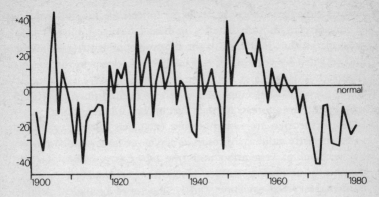

Sahel rainfall, percentage above and below normal. The recent 16-year drought was the worst on record, but precipitous fluctuations are always on the cards. Source: Sharon Nicholson, 1982

temperatures and currents with atmospheric changes spanning thousands of miles. Perhaps the best known of these are the El Niño events in the Pacific, in which the low-pressure zone normally parked over Indonesia migrates eastwards, bringing droughts to Australia, and warm currents to the coasts of South America. El Niño appears to have a fairly strong correlation with droughts in Southern and Eastern Africa, though these rarely last more than a year or two.

The persistence of abnormal conditions in the Sahel points to the possibility of some kind of positive feedback acting to reinforce climatic shifts – but controversy still rules over the mechanism. One theory proposes that drought, by reducing the vegetation cover, increases the albedo or reflectance of the land surface. As a result, more solar radiation bounces straight back into space, upward convection is reduced, and subsidence of air is reinforced. Subsiding air, as we have seen, tends to heat up, and rainfall is suppressed. A second theory builds on the fact that atmospheric dust increases in droughts. The layer of dust aloft absorbs radiation and becomes warmer than the air below. This prevents warm air below from rising and reduces convection. A third possibility is that the reduced vegetation of drought periods cuts the availability of biological particles around which moisture may condense or freeze, or indeed the overall amount of moisture in the air. Studies in Brazil have established that most of the rain that falls in the interior is moisture transpired and recycled by vegetation, especially trees, closer to the coasts. Although this has not yet been confirmed in West Africa, it seems plausible that much of the Sahel's rainfall may have been recycled up from forests on the coast of the Gulf of Guinea.

These theories are not mutually exclusive; they may all be valid or partly valid. Most of them have one feature in common: the removal of vegetation and tree cover is a central factor, whether in increasing surface albedo and atmospheric dust, or reducing recycled moisture and ice nuclei of biological origin. Drought itself reduces vegetation cover and creates the conditions for its own continuation.

Fortunately, the feedback mechanism in the Sahel, if there is one, is not perfect. If it were, droughts would last forever. External fluctuations, if they are big enough, can break the vicious circle and restore a healthier level of rainfall and veg-etation. At that point a virtuous circle can set in, in which better vegetation cover increases the likelihood of rain.

The vacillation in the Sahel between long dry spells and long wet spells has brought another problem. For almost two decades before the present prolonged drought set in in 1968, rains had been well above the trend. Over those years farmers who could not find land in their home area pushed the frontier of cultivation northward towards the desert, into what had been pasture, while the pastoralists themselves moved into what was once semi-desert scrub. Both groups were edging themselves out along a limb which could hold them only so long as the rains were good. When the long decline set in from 1968, these groups found themselves exposed and vulnerable, grazing areas that could not sustain grazing, or growing crops on fields that ought to have been pasture. The retreating rains left some destitute, and forced others back south again, into a hinterland that had meanwhile filled up with more people than it could sustain.

Given the climatic realities of Africa, crop failures due to drought or ill-timed dry spells in the growing season are not exceptional, once-in-a-lifetime mishaps. In all but the most humid areas, they come along often enough to be a basic feature of life, and to have a profound effect on attitudes to size of family and herds of livestock, and to investment or unfamiliar innovation. When the climate is so unavoidably risky, all other risks have to be studiously avoided.

THE SENSITIVE EARTH

Africa's soils are no less problematic than her climate (see map, p. 349). Only 19 per cent of the soils of the continent have no inherent fertility limitations – the lowest proportion of any region of the developing world save South-West Asia. The rest of Asia enjoys almost twice Africa's proportion of fertile soils.

Africa has only limited areas of recent mountain-building, which provides parent material for richer soils. Her rocks are, for the most part, the heavily weathered remnants of the Earth's earliest geological period, the pre-Cambrian, between 570 million and 4,000 million years old. The parent material, mainly granites and gneiss, weathers into relatively large grains, so that Africa's soils tend to be coarser than Asia's or Latin America's, with a lower content of fine clay particles.

The climate, too, has a powerful impact on soil. The heavy rainfall leaches out the soluble nutrients. High temperatures break down organic matter more rapidly and inhibit the work of bacteria that fix nitrogen from the air. As a result, Africa's soils are among the least fertile in the world. They are often low on nitrogen, which builds leaves, and on phosphorus, essential for root growth. The predominance of coarse particles and the lack of organic matter make for soils that are poor at holding water or nutrients.

Soil erosion is due to the interplay of three main factors: the slope and structure of the soil; the erosive power of wind and rain; and, intervening between the two, the amount of protective vegetation cover.

Africa is less mountainous than other continents – only 43 per cent of its land has more than an 8 per cent slope. But the low proportion of clay and organic matter in many African soils makes them particularly susceptible to erosion. Both help soil

particles to hold together or aggregate into larger clumps which
are less easily washed or blown away. Walk across a tilled field
in most parts of Africa, and your boots will sink in as if into sand
dunes. As one African agriculturalist put it, 'Our soils are easy
to work – and easy to lose.'

Water erosion is the most widespread threat. Rainfall in
tropical areas generally is highly erosive. Rain causes erosion
when it falls at more than 25mm an hour. Only 5 per cent of
rainfall in temperate areas is erosive. The proportion in tropical
areas is around 40 per cent – much of that at even higher and
more destructive velocities. Downbursts of 100–150mm an hour
are not uncommon – as much rain as New York gets in an
average month.

What protects Africa's vulnerable soils from her winds and
her highly erosive rainfall is vegetation. Trees, shrubs and grasses
break the force of wind and of raindrops. Their root growth,
plus the activity of the earthworms and termites they foster,
creates thousands of pores and channels through which the rain
can filter into the soil. But when vegetation is removed, the soil
is exposed to the power of wind and rain. Wind can pick up and
carry away soil, or deposit sand where it damages fields and
crops. Raindrops impact on bare earth with full force, sending
up a spray of droplets laden with soil particles. The pores and
channels become clogged. The rain, instead of filtering into the
soil, sheets over it, collecting here into streams, there into raging
torrents so strong a man could not stand up in them. The load of
particles turns seasonal streams coffee-brown or brick-red with
eroded soil, giving them added corrosive power, like the sand on
sandpaper, to scour out rills and gullies. Downstream the sedi-
ment clogs waterways. Dams silt up prematurely, the efficiency
of irrigation systems is depressed, the risk of flooding increased.

The increase in erosion after removal of vegetation is spectacular. In one series of studies, the annual rate of soil loss from
under forest was virtually nil – a mere 30–200 kilos per hectare.
Under crops the losses ranged up to 90 tonnes each year. From
bare soil (common on cropland at the beginning of the rains) the
rate of soil loss ran from 10 tonnes a hectare up to a massive 170

tonnes. At the latter rate, soil that took a century to form would wash away in a single year.

What this means is that maintenance of a healthy cover of vegetation is crucial to the stability of Africa's environment. Weakening and removal of vegetation is fatal: and that, as we shall see, is precisely what has been happening.

Overlaid on these more general problems are difficulties specific to certain regions. In many areas with silty soils, crusts form easily on the surface, sealing it against rainwater and creating effective drought in the soil, whatever the rainfall. Again, organic matter and clay protect against crusting, but Africa is short on both.

More than a third of Africa's land area is under desert and rocky soils too dry or shallow to grow crops. On the fringes of the desert are the sandy soils, making up one fifth of the area. These have an extra-low clay content. What rainfall there is seeps quickly down, as if through a sieve. Soil temperatures are extremely high and can easily wilt emerging seedlings. Evapotranspiration – the loss of moisture through plants' leaves – is very high. Wind erosion is a serious hazard in semi-arid areas. Sandstorms can bury infant crops under two centimetres of sand.

The humid and sub-humid areas have their own set of drawbacks. Rainfall is heavier here, so nutrients are even more rapidly leached out of the topsoil. The soil is more acid, and insoluble compounds remain close to the surface: aluminium oxide, which is toxic to plant growth, and iron oxides, which give so many of Africa's soils their characteristic rusty hues. Frequently the oxides crystallize into a cement-like hardpan which roots cannot penetrate. Where the topsoil is washed away this hardpan forms an impenetrable rocky surface, and cultivation becomes impossible.

Africa has areas of very fertile soil. Most are in the highlands – the only extensive area of recent mountain formation, stretching from Northern Ethiopia down the Rift Valley as far as Northern Tanzania. The highlands pay for their rich soils with steeper slopes that increase the risk of erosion exponentially. Another set of potentially fertile soils are the poorly drained and dark clay soils. These suffer from waterlogging in the wet season. In the dry season, when the residual moisture could be used to grow crops, the dark clay soils crack and harden so they become unworkable with traditional technologies. In Ethiopia, some of the best soils are in the valley bottoms, which collect the debris washed down from the hills. Most of them are grossly underused: in the wet season they become lakes, while in the dry they graze cattle.

TRADITIONAL ADAPTATIONS

All in all, the soil map of Africa, when overlaid by the harsh realities of her climate, presents the most daunting challenge to agriculture of any continent.

No other continent suffers the same degree of separation of agriculture and livestock. No other continent has such a high proportion of soils that are so infertile and so easily degraded. No other continent has a climate of such fearful unpredictability.

These factors – made more potent still by their combination – have severely handicapped African agriculture. Outside a few favoured areas, it has been more difficult to produce an agricultural surplus, over and above the needs of survival. When surplus was produced in one year, or one run of years, it was often wiped out by one or more years of drought.

Africans have adapted to their difficult circumstances as best they can. The dominant form of agriculture – common all over the world wherever population densities are low – has been shifting cultivation. In broad outline, farmers clear an area of bush or forest with axe or cutlass, and burn the debris. This burning wastes much of the nitrogen in leafy vegetation – but it reduces the acidity of the soil, often a serious problem in humid areas, and destroys weed seeds, insects and plant diseases. Food crops are then cultivated for one to three years. The first crop draws on soil nitrogen and other nutrients built up in the fallow years, but after that yields decline rapidly by 10–20 per cent for each successive crop, until the return is no longer worth the labour invested.

The field is then abandoned. Grasses and other plants re-invade the plot, and rain and soil bacteria restore lost nitrogen. A crucial role is played by shrubs and trees. Their deep roots reach down to the soil layers where rain may have leached minerals. The falling leaf litter restores these nutrients to the topsoil, along with organic matter. The fallow areas also act as soil and water traps, collecting much of the rainwater that washes off the cultivated plots and the earth it carries with it. Underground aquifers are replenished, and the soil which a plot loses under cultivation, it often regains under fallow. Fallow areas and virgin forests are also used as sources of fuelwood, grazing, and other products.

Shifting cultivation is essentially a no-input form of agriculture. It relies on nature to restore the nutrients removed in crops, and to provide fuel. So long as nature is given time enough to do so, the system is stable and sustainable. It is still the dominant mode of farming in all but the most densely populated zones of Africa. Accurate estimates of its extent are hard to come by, but the FAO survey of tropical forest resources found that in 1981 there were some 166 million hectares of woody fallows in tropical Africa, against 139 million hectares of currently cultivated land – fallows, in other words, represented some 54 per cent of the whole area available for farming. But the regional variations were very wide. In West and Central Africa, 70–75 per cent of the farm area was under fallow. In the Sahel, only 43 per cent was fallow, and in more densely populated Nigeria, Kenya and Uganda, only 20–30 per cent. In overcrowded Rwanda and Burundi, fallowing had almost disappeared.

Within the areas that are being cropped at any one time, the peasants' main solution to the problems of erosion and unstable climate has been intercropping: growing two or more crops, with complementary needs for light, soil or water, intermingled in the same field. Intercropping has been scorned and discouraged as primitive and messy, but as we shall see (p. 108), recent research shows that it yields more total output from the same area than when crops are grown alone. This yield is also more reliable from year to year: an intercropped field will produce a lot more than a sole crop in dry years. And in all years, two or three crops with different maturity periods keep the soil covered for longer and protect it against wind and rain.

INTENSIFICATION, AFRICAN STYLE

It is a general rule of agricultural development that, as population density increases, there is a gradual move from extensive to more intensive forms of farming. Instead of moving the fields, the same plot is cultivated most or all of the time, and the nutrients taken out by crops are put back in the form of fertilizers, manure, crop residues, or the use of nitrogen-fixing bacteria in legumes like beans or peas.

The older inhabited centres of the Near East and Far East made the transition to intensive farming millennia ago. There the predominant form of settled agriculture is irrigation. The soil's fertility is maintained by blue-green algae in the water, or by nutrient-rich silt washed down from young mountain zones. Europe made the shift gradually, from the Middle Ages on, with more and more sophisticated crop rotations involving legumes, and the integration of livestock and arable farming.

Africa's transition is more problematic than in any other area. The Asian option of irrigation is much more difficult. The African landscape, outside East Africa, is generally flat, and her rivers are highly variable in their flow. Surface waters are less common, and shallow groundwater is harder to find. The opportunities for cheap and simple irrigation are limited.

The European path to intensification, mixed farming, is also problematic. Trypanosomiasis blocks the integration of livestock rearing and cultivation over much of the continent, so manure for fertilizer is in short supply. Chemical fertilizers, which have largely replaced crop rotation and manure in Europe, are hard to come by given the poverty of African peasants and their governments.

Time is another luxury Africa does not possess. In other regions the shift from extensive to intensive farming took centuries. Given Africa's explosive rate of population growth, currently 3.2 per cent a year, the shift needs to take place in two or three decades.

The final problem is posed by Africa's sensitive environment, which is much less tolerant of reduced tree and plant cover than most of Asia and Europe. Soil conservation works are needed at much lower population densities, before other aspects of intensification become necessary. The cost of moving over into settled agriculture may well be higher in Africa than elsewhere, and that cost may well be acting as a barrier.

Over most of Africa, there is still enough land for it to be cheaper and easier, in the short term, for farmers to move their plots every two or three years. *Population density has not yet reached the level where intensive farming in unavoidable. But it has reached the level where massive ecological damage can occur if traditional methods go on being used.* Most of Africa is presently caught in the gap between those two critical densities. In northern Europe's stabler environment, the gap probably does not exist.

There are pockets of intensive agriculture in Africa, mainly in the highlands of East Africa and in South Eastern Nigeria. We shall look at these in more detail in Chapter 4. But most African farmers are still using the techniques of shifting cultivation, increasingly under conditions that make those techniques destructive.

The cropped area has expanded, and the fallow period has been progressively reduced, in many areas to the point where it can no longer restore soil fertility or yield sufficient wood or grazing. Trees no longer have time to grow to maturity, sometimes not even to set seed. Nutrients and organic matter are no longer returned naturally to the topsoil, nor do farmers restore them artificially by chemical or organic fertilizers. Except in severe droughts, livestock numbers grow in parallel with human numbers. The growing herds compact the soil, stripping cropped and fallow areas of edible vegetation, chomping tree seedlings down to ground level. Fuelwood needs grow in line with population too, and press on a reduced tree stock. Crop residues and dung, so desperately needed to maintain soil fertility and structure, are increasingly burned as fuel.

A series of vicious spirals sets in. *All pressures push together to reduce the protective cover of vegetation, exposing the fragile soil to the battering of tropical downpours and winds.* The reduction in organic matter makes the soil even more easily eroded or crusted over. Less rainfall filters in to feed crops and pastures, more runs off destructively. No longer recaptured by surrounding fallow areas, more water and soil flush away down river courses. Soil depth declines, so plants cannot root deeply. Soil fertility declines as it is the most fertile topsoil that is removed first. Depleted of organic matter, the soil is less able to hold water. *Again, all pressures push to create a shortage of water in the soil, even in years of good rains*. Starved of moisture and nutrients, plant and tree growth is reduced even further, giving the spiral yet another twist.

Drought tightens the spiral. Vegetation cover and organic matter are reduced, shrubs and eventually trees die, or are felled for sale as wood or charcoal by farmers in need of cash to make up their deficits of grain. People and animals migrate to areas less affected by drought, increasing the pressure there, too.

Conversely, overcultivation may intensify and prolong droughts. There is no doubt that it has increased agriculture's *vulnerability* to drought, since less of the rain that does fall filters into and remains in the topsoil. But it is possible, too, that the processes may lead to lowered rainfall. As we have seen, most of the suggested feedback mechanisms for prolonged drought in the Sahel involve the reduction of tree and vegetation cover. Where human beings and their animals contribute to these processes, they can prolong the incidence of drought. And if much of the moisture that falls inland in West African Sahel is recycled moisture, relayed by trees stretching down as far as the coast of the Gulf of Guinea, the heavy rate of deforestation in the coastal countries – running 3–4 per cent a year – may well have contributed to the worst prolonged drought on record in the Sahel.

It is easy, with hindsight, to blame Africa's farmers for the ecological disaster that is in progress. Yet Europe's farmers cleared vast areas of forests and woodland for agriculture and fuel without much thought for the consequences. Only the gentler climate and more stable soils saved them from a similar catastrophe. In many environmental issues the catastrophe actually arrived before action was taken: America's dustbowl, the drastic decline of some whale species, the death of forests killed by acid rain. The fact is that the peculiar dangers in Africa were not widely appreciated until they actually materialized. Experience is teaching Africa's leaders and farmers in the school of hard knocks, but a great deal of damage has already been done, and the task of repairing it is that much harder.

AFRICAN CONSTRAINTS

Climate

- *Rainfall variability:*
Spells of drought in growing
  period
Years of drought
In Sahel: prolonged droughts

- *Rainfall in storms:*
Leaching of nutrients
High proportion of erosive rain
High water loss through run-off

- *Alternation of dry and wet seasons:*
Death of annual vegetation
Soils bare at first storms

- *High temperatures:*
High evapotranspiration
High rate of decomposition of
  organic matter

Soil

- *Low clay content* plus *Low organic
  matter content:*
High erodibility
Crusting
Poor water-holding capacity
Poor nutrient-holding capacity
Low fertility – low nitrogen and
  phosphorus

- *Humid zone:*
High acidity
Aluminium toxicity
Hardpans of iron oxides

- *Semi-arid zone:*
High soil temperature
Sandstorm risk
Sandy soils – poor at water-
  holding

- *Good soils associated with:*
Steep slopes therefore high erosion
  risks
Poor drainage and waterlogging

Biological

- *Rapid weed growth*
- *High disease and pest incidence:*
Low labour productivity
Crop losses
Low animal productivity

- *Trypanosomiasis:*
Blockage of integrated livestock
  and arable farming – lack of
  animal power – labour shortage

Water

- *High seasonal variation of rivers*
- *Low surface water availability and high evaporation*
- *Few sources of shallow groundwater*
- *Flat topography*
- *Difficulty of cheap irrigation*

Social and Economic Constraints

- *Small farm size*
- *Fragmentation of plots*
- *Shortage of cash for investment*
- *Risk avoidance:*
Reluctance to increase exposure to risks
Tendency to overstock with livestock
High birth rate

- *Uncertain tenure:*
Communal and/or state ownership of land, forests, water points, rangeland
Reluctance to invest in long-term improvements
'Tragedy of the commons'

- *Labour shortage:*
Due to:
Lack of animal power
Male labour migration
Women's excess burdens
  70% of food production, 100% of processing
  Child rearing
  Wood and water gathering

Leading to:
Late ploughing and planting
Late weeding, late harvesting
Lowered yields

- *Women's status:*
Lack of power to obtain credit, plan cropping
Insecurity of tenure: no incentive to improve land

3 Why things go wrong in Africa: the record of failure

No one who has travelled much can fail to notice how things tend to go wrong in Africa. There is hardly a hotel room without its stock of candles against inevitable power cuts. Projects are paralysed as jeeps are mothballed for months, waiting for spares. Graveyards of immobilized, rusting bulldozers or mechanical shovels are a common wayside spectacle. Communications are unreliable. The post and telephone system in Nigeria is so erratic that large companies have to depend on radio links and special courier services.

The environment does not help. Dirt roads compress into ridges like a corrugated tin roof, shaking vehicles gradually to bits. Tyres wear out faster – in eight days of up-country touring in Ethiopia I had four punctures. Everything wears out faster. In the humid areas rusts and moulds eat into buildings and equipment. In dry months dust and sand coat surfaces, creep into crevices, and jam moving parts.

Projects, policies and programmes are as breakdown-prone as equipment. Major aid agencies like the World Bank and USAID are increasingly frank about the disastrous performance of aid in Africa. In 1985 the World Bank looked back on the performance of more than a thousand of its projects that it had evaluated over the previous decade. In developing countries as a whole, an average of 12 per cent failed to achieve their objectives or had an uncertain outcome. But in the two sub-Saharan African regions, the failure rate was much higher – 18 per cent in West Africa, and no less than 24 per cent in East Africa. Average economic rates of return on projects were also lowest in the two African regions – little more than half the South Asian average in the case of East Africa, and only two thirds in West Africa. Agriculture, the most crucial function for Africa, fared worse than any other sector. In West Africa almost one agricultural project in three was judged a failure; in East Africa just over a half. Against this, in South Asia only one project in twenty failed.

Donors have also paid close attention to 'institution building' – strengthening the staffing and management of government and non-government organizations – but here the outcome has been even more dubious. World Bank measures to strengthen institutions were substantially successful in only 28 per cent of cases in sub-Saharan Africa – against almost half in other regions. USAID's record was no better – out of 183 African projects involving institution building, positive results outweighed the negative only half the time. Indeed, in over two projects out of every five, the negative results outweighed the positive.

There are good reasons for believing that the true picture is even worse than these depressing figures suggest. Most evaluations are made at the end of a project, which typically lasts from three to five years. But it is in later years, after the foreign funding has been withdrawn, that the most serious problems arise. In 1985 the World Bank looked at the longer-term impact of 25 agricultural projects – 17 of them in Africa. All of them seemed successful at the end-of-project audit, producing economic returns of 10 per cent or more. But after five or ten years, 13 of these projects had not sustained their benefits. All 13 were in Africa. Their average rate of return was now estimated at less than 3 per cent, and in two cases was negative.

In a number of fields that are central to African environment and agriculture, the record is nothing less than appalling. World Bank African livestock projects reviewed in 1985 had *negative* economic returns averaging minus 2 per cent, against an average *gain* for other regions of 11 per cent. The experience in irrigation was so bad the Bank considered halting all loans to new large-scale irrigation projects in Africa. A USAID review of forestry projects concluded in 1984 that no large-scale plantations or even village woodlots in the sensitive Sahel area had been successful.

All these findings relate to *aid* projects. There are no systematic

surveys of the effectiveness of *government* programmes in Africa,
but it is safe to assume that the picture is at least as bad. For
one thing, in 1984, aid amounted to two thirds of the gross
domestic investment of the 25 low-income countries, and there
are few programmes where aid is not involved in some way. The
record of declining incomes and food production per person over
the past fifteen to twenty years is a clear pointer to widespread
failure.

THE COLONIAL IMPACT

The aim of this book is to show how things can go right in
Africa, and do go right when the right things are done in the
right way. But an important premise is to understand why so
many things have gone wrong in the past. The colonial past,
and its effect on the successor African states, bears a heavy
responsibility. Colonialism was everywhere disruptive and
destructive of traditional societies, economies and established
authorities – but nowhere with more far-reaching effects than in
Africa.

In the Americas and Australasia, colonialism took the form of
colonization. European settlers moved in, expropriated the land,
and exterminated most of the natives. The survivors were marginalized and confined to enclaves, sizeable only in the Andes.
The new states were run by and for Europeans. Problems of
culture conflict or technology transfer were minimized.

In Asia, urban civilization and centralized bureaucratic states
based on the surplus of irrigated agriculture were centuries,
often millennia, old when the colonial powers arrived on the
scene. With the long tradition of literacy, the European rulers
found it easy to recruit and train junior civil servants locally.
Colonial rule lasted for a century and a half or more, long
enough to have some beneficial as well as negative effects: the
creation of a cadre of bureaucrats like the Indian civil service,
and more basic infrastructure in roads, railways and major
irrigation works. Asia embarked on independence better prepared than Africa.

Africa itself was long protected from direct colonization by diseases like malaria and yellow fever, with death tolls estimated at 50 per cent a year among Europeans foolhardy enough to live inland in West Africa. For three centuries the slave trade exerted a pernicious influence, intensifying inter-tribal conflicts, especially in West Africa. But it was not until the mid-nineteenth century, after the discovery of quinine as a malaria cure, that Europeans moved inland in any numbers, and not until the 1880s that large tracts of land began to be appropriated in the scramble for Africa unleashed by the Congress of Berlin in 1884.

Cooler areas like Southern Africa and the highlands of East Africa had the misfortune of a climate that was agreeable to Europeans, and attracted large numbers of settlers. The communal landowning pattern, the absence of written land titles, and the large areas of fallow made expropriation simpler. In Kenya, Zambia, Zimbabwe, Angola, Mozambique, Swaziland and South Africa, whites seized the best land by force, while local tribes became landless labourers, or were forced to move into marginal areas which were much more easily degraded.

The cultural distance between conquerors and conquered was far wider than in Asia. Africa's political and social organization was primarily determined by her low population density and the system of shifting cultivation which it permitted. Shifting cultivation does not require a complex bureaucratic state, with a literate ruling class. In some areas there were larger, more complex states, usually based on the regulation of long-distance trade, but with a fairly narrow urban base.

The Europeans imposed a bureaucratic, literate state on peoples mostly at a pre-literate, tribal level of social organization. The colonial boundaries were dictated by great-power relations and the location of trading posts and military expeditions. From an African point of view they were arbitrary, often cutting across tribal boundaries, nearly always including disparate and sometimes hostile tribes within the same colony. Traditional authorities were either suppressed and replaced by administrators (the French direct rule approach) or converted into collaborators and stooges (the British style of indirect rule used in West Africa).

Over much of Africa the ability of African communities to

control their own destinies or their environment was undermined. Fallow lands essential to soil fertility, common forests that provided wild game, fruits and fuel, seasonal pastures left empty for much of the year, were nationalized, or expropriated for European settlers or plantations. A whole series of policies regulated Africans' relations with their own environment. The 1935 forestry code of French West Africa created state forests, protected against the 'encroachment' of communities for whom they had, from time immemorial, constituted a vital resource. Even outside the state forests, many tree species were protected, so that villagers were supposed to get permission to harvest fuelwood from their own trees.

In many countries the wildlife that had provided extra protein became the preserve of the big white hunter. The African could no longer benefit from, or control, the wildlife of his own environment. He became, as Norwegian historian Helge Kjekshus put it, an outlaw in his own land. Pastoralists suffered a similar disenfranchisement. Rangeland, though covered by no official titles, was usually controlled by particular clans or tribes. Colonial legislation in many cases nationalized the pastures. New wells provided by governments, and therefore free for all, effectively threw the rangelands open to all comers and destroyed traditional controls.

The economies of the African colonies were systematically exploited for the benefit of the mother country, creating a dependence that still persists today. Mines, plantations and cash cropping schemes were developed as sources of primary commodities. Traditional rural industries like weaving, pottery and metalworking were undermined or wiped out by imports of manufactures.

The mines and plantations required labour, so means had to be found to compel African farmers to work in them. In the French colonies Africans were forced to do compulsory but paid labour for commercial companies. The Sahelian countries were milked as labour reserves for the plantations of the coast. In the British colonies poll taxes, which had to be paid in cash, forced Africans to work for European employers. Plantations required mainly seasonal labour – but the season often coincided with the time when village food crops needed most attention. Mines required permanent labour, with the miner returning home for only a few days each year. Male labour migration became an institution that persists even today, leaving many African wives as grass widows, farmers and housewives combined.

AN INSECURE INDEPENDENCE

Sub-Saharan Africa was the last region to be colonized and the last to achieve independence. Europe abandoned Africa with little infrastructure in the way of roads and railways and a very limited pool of educated or technically qualified manpower.

The economic distortions of colonialism persisted after independence. Cash crops and minerals remained the chief source of foreign exchange and a major source of government revenue. The switchback fluctuations of commodity prices wreaked havoc with trade balances and government budgets, making coherent long-term planning and spending impossible. The long-term price decline of most commodities compared with manufactures and oil led to deepening debt, import cuts and budgetary restraint. As local industry failed to take off, African countries remained heavily dependent on imports for all kinds of machinery and inputs for industry and agriculture.

The arbitrary colonial frontiers, adopted by the new black rulers, created their own set of problems. Africa has more landlocked countries without access to the sea than any other region: fourteen in all, twice as many as Asia and Latin America put together. Because of transport costs, these countries, often hundreds of miles inland, pay far more for imports and earn less for their exports. In West Africa, colonial boundaries have created an artificial separation of humid coastal areas and semi-arid inland areas, which should logically be interdependent.

The frontiers were as problematic for what they included – the diversity of tribal and linguistic groups – as for what they excluded. Countries with many small tribes, like Tanzania, or a single overwhelmingly dominant tribe, like Swaziland, fared best. In countries with several large ethnic groups, tribes competed for power and some of the unsuccessful contenders launched secessionist wars. Serious ethnic or regional conflicts arose in Burundi, Rwanda, Zaire and Nigeria, and are still in progress in Chad, Sudan, Ethiopia and Uganda. Border conflicts occurred between Somalia and Ethiopia, Mali and Burkina Faso. There were protracted wars of independence in Kenya, Angola, Mozambique, Guinea Bissau, Zimbabwe and Namibia, while South African destabilization policies prolonged the civil wars in Angola and Mozambique. Altogether, seventeen countries, with 70 per cent of the population of sub-Saharan Africa, have suffered severe and prolonged conflicts since the mid-1950s.

As a result Africa, with less than 10 per cent of the world's population, harbours almost half the world's 10 million refugees. Another effect has been to divert resources into military expenditure that could have been used for development. Sub-Saharan Africa spent 3.2 per cent of her gross national product on military expenditure in 1982. This is a lower proportion than South Asia (3.5 per cent) or the Far East (6.4 per cent) but far higher than Latin America's modest 1.4 per cent. If Africa spent the Latin American proportion, an additional $3.5 billion a year would have been available for productive purposes in 1982 – almost twice the total expenditure on health in that year. Equally serious is the general feeling of insecurity that military conflict creates. Food production suffers as farms are abandoned, or planting, weeding and harvesting are disrupted. Investment becomes even more risky.

Most African countries experimented with Western-style democracy in the early years of independence, but the narrow social base of most governments and the problems arising from ethnic and emerging class conflicts soon put an end to most Western-modelled constitutions. By 1985 more than three quarters of all African governments were either military or one-party regimes.

Authoritarian, top-down styles of government became the norm. The gap between the centralized modern state and the village is almost as wide as in colonial days. The local powers that existed in pre-colonial times have not been restored in most countries. The business of government is conducted in writing, and the majority of adults are still illiterate – 61 per cent in

1980, far more in rural areas, against only 38 per cent in Asia and 20 per cent in Latin America. Africa is the only region in the world where government is conducted in languages that, as far as the mass of the population are concerned, are foreign. This makes it doubly difficult for illiterate peasants who have little command of the official language to get across their demands and grievances. The information traffic is one way, from the centre to the periphery.

THE FAVOURED CITIES

In Asia and in Latin America, there are large landowning interests with enough economic and political muscle to make sure that the interests of agriculture are pushed. In most of Africa, nearly all farmers are subsistence smallholders. Communal land ownership meant that land could not be bought and sold. Substantial wealth in Africa could only be accumulated in urban areas. The cities became the focus of ambition and competition.

The groups that did have influence on government were predominantly urban. Largest of these were the public sector – civil servants, junior public employees organized in trade unions, and of course the military. Next came urban business interests – property speculators, importers and traders, owners and managers of infant industries – often with close patronage and family links to top civil servants and politicians. Finally there were the urban masses, housed in self-built shanty towns, employed in factories and petty trades from street hawking and whoring to gardening and night-watching. The masses were not much better organized than the peasants – but they had the advantage of living close together, and held the ultimate weapon of all city dwellers, the power to riot if food prices or bus fares rose too high. All in all the balance of power lay overwhelmingly in favour of the cities.

The upshot was that government policy in almost every sphere was framed to suit urban interests, producing what British development expert Michael Lipton has called urban bias. The model of development pursued was to develop cities and factories. Urban areas swallowed the bulk of spending on services from paved roads to electricity, from education to piped water and hospitals. As late as 1983, for example, 61 per cent of town dwellers in Africa had access to clean water, but only 26 per cent of rural people. City incomes are far higher than rural ones in Africa, mainly because the civil service overpays itself in most countries. Average incomes in the service sector in Africa were more than seven times higher than in agriculture in 1980, whereas in Asia and Latin America they were only three and a half times higher.

Industrialization was a prime goal. But instead of following the export-oriented strategy that brought such rapid results in the Far East, the aim was import substitution to produce at home the manufactures that had been imported. This meant beginning with consumer goods – beer and soft drinks factories were often the first to start up – following on later in larger countries with industries assembling consumer durables from imported components. The ostensible aim of all this was to save on foreign exchange – but the effect was often the reverse. The market for machinery was so small that a capital goods sector did not develop: industry continued to rely on imports for all its machinery and spare parts, and often for raw materials inputs like plastics or steel. The economics of these operations were frequently questionable. One African country, for example, set up a polyester yarn factory. The petrochemical inputs and all the machinery had to be imported. The cost of making the yarn at home was one third more in foreign exchange alone than it would have cost to import the yarn.

Given a parasitic bureaucracy and an inefficient industry, there was only one genuinely productive sector in African economies – agriculture. Farmers were milked to provide the funds for government and industry. In some countries like Upper Volta (now Burkina Faso) they paid taxes, and received no services in return. In most countries government monopolies bought up their cash crops, and paid them far less than they were worth on the world market, pocketing the difference for government revenue. Farmers in the main coffee, cocoa and cotton exporting countries in Africa received only about half the export value of their produce in 1984.

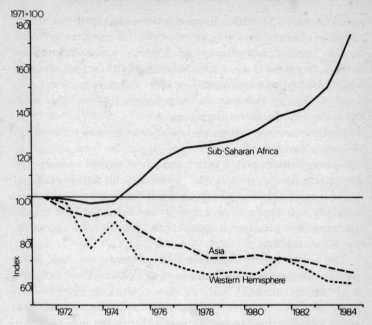

1971=100

Real effective exchange rates 1971–85. Source: World Bank

The whole trade and exchange policy was framed to favour urban interests. Currencies were not allowed to float or fixed at realistic levels, but overvalued, often by 100 per cent or more. In 1984, despite Africa's crippling debt, her worsening terms of trade, and her lagging exports, the real effective exchange rates of African currencies were 60 per cent up on 1971. The exchange rates for Asian and Latin American countries, with stronger economies, were more than 30 per cent down over the same period. Overvalued exchange rates mean that imports of the machinery and raw materials of import-substituting industries are artificially cheap. So are imports of food, and of whatever consumer goods traders could get licences to import. Urban interests benefited almost exclusively.

The peasants suffered. Cheap food imports, plus food aid, kept food prices – and therefore the price farmers got for their produce – low. The imports were mainly wheat and rice. Many urban consumers came to prefer them for taste and convenience, so the demand for locally grown staples – maize and cassava, millet or sorghum – was further depressed. Overvalued exchange rates also reduced the value, in local currency, of cash crop exports, further reducing farmers' incomes. They also made other exports uncompetitive abroad.

Keeping an overvalued exchange rate does not increase a country's foreign exchange receipts – it reduces them. Almost all African countries operate with a scarcity of foreign exchange. As the price is fixed artificially low, the demand for foreign exchange greatly exceeds the supply. Complex bureaucratic procedures ration it out: applications are long and laborious, the process absorbs scarce qualified manpower, and provides rich opportunities for corruption.

The whole strategy has been a disastrous failure. Low prices paid to farmers have kept food production low. They have also kept farmers' incomes low, so they could not afford to buy manufactures. The market for industry has been restricted to the small formal urban sector. Import substitution led, paradoxically, to continued import dependence. Exchange regulations and foreign exchange crises interrupt supplies of spare parts and raw materials, so factories work at only a fraction of their capacity. African industry has been crippled by the very policies designed to foster it.

Yet the trap is a hard one to break out of. The higher incomes and better services of cities have acted as a magnet for the rural poor. In 1950 only 33 million Africans – 15 per cent of the total population – lived in cities. Thirty years later the numbers had quadrupled, to 137 million, 29 per cent of the total. In 1980 Africa's cities were growing much faster than those of any other region, at 5.3 per cent a year. The larger the urban share of the population, the harder it becomes, politically, to shift the balance in favour of farmers.

Added to the urban bias has been a Western bias, a tendency to overvalue Western technologies and approaches and devalue African ones. The African peasants' ways of shifting cultivation, untidy mixtures of crops in the fields, annual burning of cleared bush and rangeland, were alien to European eyes. Most colonial administrators could not make an imaginative leap to understand why African peasants and pastoralists did the things they did, and condemned them out of hand. Most viewed their task, as colonial administrators in East Africa put it, as 'changing a disease-ridden and backward horde of savages into a disciplined and prosperous community', or 'saving the African from himself'.

The educated elite maintained similar attitudes to traditional farming after independence. The first teachers of agriculture at African universities were Westerners, visiting experts were Westerners, and the slowly expanding cadre of African experts perfected their training at Western universities. Since Western agriculture was clearly more productive than African it seemed logical that European technologies and approaches should be introduced: full-scale mechanization, chemical fertilizers, mono-cropping. And as African smallholders seemed to resist these innovations, many governments came to see vast state farms or large-scale commercial holdings as the answer to their food problems. The Western bias blinkered programme designers and policy makers against the special character of African soils and climates, and the value of traditional ways of dealing with them.

Finally, the new African states for the most part perpetuated the state ownership of wildlife, rangelands and natural forest. All these resources had previously been used by well-defined groups. As population density increased, there is every reason to assume that local communities would have begun to manage their resources and control their use. State ownership took away the possibility of *community* control, but as budgets were tight, there was no way to employ enough wardens, foresters or range managers to ensure effective *state* control. It was now in the interests of every individual to take what they could from the commons before anyone else did. What American anthropologist Garrett Hardin has called the tragedy of the commons came into operation. Every household stood to gain the full benefit every time they decided to lop an extra branch for fuel, or to stock one more animal: but they stood to lose only a fraction of the ecological damage caused by each act.

THE PEASANT'S REALITIES

Far below the political plane lies the African farmer, usually neglected, sometimes seriously exploited. The social and economic realities of peasants' lives make development efforts in Africa more difficult than in any other region.

The first problem is the sheer variety of tribes, economies and farming systems. The low population density fostered schisms and separations, and Africa's four or five main ethnic families segmented into literally hundreds of sub-groups with distinctive languages and cultures. Nigeria alone has some 300 local languages. Criss-crossing this human diversity is the diversity of habitats: even within the limited area of one small tribe, there may be dramatic differences according to altitude, soil, steepness of slope, position on the slope, height of water table, and so on. The complex patchwork fabric makes it impossible to impose nationwide solutions that will work in all circumstances.

The dominant system of landownership before colonial times was communal. As usual in areas of shifting cultivation all over the world, land was owned by the whole community. Most of it was fallow, used communally for grazing animals, gathering wood, herbs, honey and other products. The area under crops shifted every two or three years. It was allocated to individual families by village chiefs, with larger families getting larger holdings. They did not *own* the land they were given: but they owned its produce as well as the produce of any trees they planted. To ensure fairness, families would generally be given a number of holdings of different quality. If you ask Africans to show you where their farm is, they will point in several directions: one Ethiopian highlander I came across had separate plots in the steep land under the escarpment, in the gentle middle slopes, and in the valley bottom, waterlogged in the wet season and used as pasture in the dry. Plots are often widely scattered, with long walks between them taking up much of the time that should be used for cultivation.

The basic agrarian pattern – like the technology of shifting agriculture – was fair, and well adapted to the situation. But as population densities have increased it has come under pressure.

In most countries formal ownership of the land is now vested in the state, but individual families tend to be given more or less permanent usage rights over certain areas, which can be passed on to children, or even, in some cases, rented out. In a few countries, like Kenya, these permanent rights have been converted into full ownership.

Most of Africa is in the transition phase between communal and individual ownership. It is a no-man's land in which farmers have permanent rights over an area, without legal title to it. Their tenure is uncertain. They cannot offer their holding as collateral for loans, because it is not fully theirs to forfeit if they default. They cannot be sure they will still be farming the same areas in ten or twenty years' time, and so they are more reluctant to invest in permanent improvements to the land, from tree-planting to soil conservation works.

The total area farmed by each family is usually small – less than three hectares on average, even in areas with abundant land, because there is not enough labour to work a larger area. Unlike Asia and Latin America, which have large pools of landless people, there is actually a shortage of labour in rural Africa. It does not bite all the year round. In the dry season, though women have to trek further for water, men seem to find time for social activities, from ceremonies to drinking parties. The labour shortage is tightest in the growing season. The lack of oxen for ploughing is one of the most crucial factors: animal power supplies only 13 per cent of power inputs to crop production in Africa, well below half the Asian level of 31 per cent. Machinery supplies only 3 per cent. Human labour supplies the other 84 per cent. Added to this are the long walks between scattered plots, the migration of able-bodied males in search of paid work, the virulent growth of weeds invading from nearby fallow areas. The labour shortage creates a damaging bottleneck around the time of the first rains. If planting or weeding are delayed, yields suffer quite dramatically. If harvesting is delayed in areas with two rainy seasons, the old crop may still be lying in the field when the first new rain arrives: it will be put away damp and damage by moulds in storage will be more serious.

The peasants' predicament is a severe handicap to investment. Because agriculture produces little surplus for sale in most areas, spare cash is in short supply. Because of the uncertain climate, there is a constant risk of crop failure, and farmers cover themselves against the risks as best they can. If they are lucky enough to have a surplus one year, they tend to store it against the next. If they have a surplus two years running, they may accumulate capital in the form of extra livestock – or an extra wife. Understandably, they avoid any action that will increase the risks they are exposed to, or reduce their home-made insurance cover.

The lowly position of women in Africa aggravates most of these problems. Women are the backbone of African agriculture. In the days when game was plentiful and inter-tribal warfare or raiding a constant hazard, men hunted and fought. Women were responsible for *collecting* plant foods in hunting and gathering days, and retained the main responsibility after food began to be *cultivated*. Women also had the job of collecting water and firewood, logically linked with food processing. By association, in many areas any carrying of heavy burdens was women's work. Men's chores were lightened as tribal warfare diminished, and game dwindled or was protected. But the old division of labour persisted and still persists. Men specialize in certain tasks in the fields – clearing bush or forest for cultivation, felling or pruning trees, ploughing or turning the land. Women, it has been estimated, do 70 per cent of hoeing and weeding, 60 per cent of harvesting, 80 per cent of transporting crops home and storing them, and 90 per cent of food processing. In many parts of Africa, especially Eastern and Southern Africa, as many as half of rural households will be headed by a single woman or the wife of a migrant worker; these women, helped by their children, do all the work on the farm. In areas where firewood is scarce, or where sources of water are distant in the dry season, women's burdens are intolerable. They may spend an average of an hour a day collecting firewood, a hour and a half pounding or grinding foods, and anything from half an hour up to six hours fetching water. Something has to give, and as water and wood are inescapable daily needs, planting, weeding and harvesting are delayed and yields suffer.

Despite women's massive share in the farm workload men remain firmly in charge of farm planning. Frequently men give priority to cash crops (whose proceeds usually go to them, to use or misuse as they see fit) rather than to subsistence food crops which merely feed the family. These are some of the complaints voiced by women in the Wedza area of Zimbabwe to sociologist Kate Truscott: 'The man makes the plan and then goes away. We do all the work.' 'The women might say "Let us plant beans or groundnuts" [the money from these crops traditionally belong to women] but he ignores us. We can't plant them, or the man would order us to go and find another man.' 'Men spend money on beer and on booking rooms to spend the weekends with prostitutes.'

In areas where men migrate more permanently for work, to mines, plantations and cities, women are not only the chief farm labourers, they are the farm managers too. Yet they have no power to spend cash on inputs or investments. In many African countries they are legal minors, not entitled to hold title to land, to enter into contracts, or to raise loans. The uncertain tenure that overshadows most African land affects women most of all. If their husband divorces them, they may find themselves landless. If he dies, they will be dependent on their eldest son for land. Thus they are even more reluctant than men to invest cash or labour in long-term improvements like soil conservation or tree-planting.

THE SOURCES OF FAILURE

Against this unfavourable backdrop, the record of failure in development becomes less surprising.[1]

The shortage of educated and trained manpower works together with tight budgets to produce a dearth of expertise. Services are understaffed in every field. Technicians to maintain

[1] The analysis that follows is based on a synthesis of reports from the World Bank, USAID and others – see Bibliography

vehicles and equipment are in short supply. All-weather roads, railways, and telephones are inadequate, and limited budgets cannot maintain them properly or expand them fast enough. This entails failures of communication and hold-ups in all rural activities, as key staff or materials fail to get through.

Africa's import dependence is a major source of difficulty, as balance of trade problems lead to regular import cuts. Large stocks of spare parts cannot be afforded: often they are not ordered until the equipment has already broken down, and take months to arrive. In times of austerity they may not be ordered at all. Even at other times, complex currency and import formalities can lead to long delays in delivering equipment and spares. When electricity supply, telephone installations, lorries or jeeps break down, there is a far-reaching domino effect as communications and deliveries of all kinds are held up. Where the imports are destined for mass use – for example, chemical fertilizers or modern contraceptives – users lose confidence and are likely to go back permanently to more reliably available traditional approaches.

One World Bank agricultural project in East Africa, for example, was dependent on imported chemical fertilizers. These were given out free in the early years, but later there was a complete ban on imports. The Bank's Bura irrigation and settlement project in arid South Eastern Kenya was crippled when newly arrived settlers from the overcrowded highlands deserted in droves, often within days of arriving, because the health and education services they had been led to expect were not there. Malaria was rife and infant mortality high. The planned malaria control programme was held up because it had no vehicles: two jeeps included in the project were held up by complex administrative procedures. Primary schools were not complete simply because a small amount of aggregate needed for the floor was not available.

Budget restrictions and recurrent budget crises are endemic to Africa's weak economies, and can severely damage projects and programmes. Donors like the World Bank complain that African governments often hold back on their share of project costs. Central departments on which projects depend for back-up may be understaffed and underfunded. When budget crises blow up, investment and equipment are the first to be slashed. Jobs are politically harder to cut back, so staff work on with inadequate resources, and much of the work they are paid to do remains undone.

Political instability compounds the effects of economic instability. Sudden changes of regime, shifts in policies and programmes, purges or rapid turnover of ministers and top-level civil servants, all play havoc with development programmes. The young, unstable state lacks legitimacy. Politicians and civil servants may put family, locality or tribe before the national interest, leading to nepotism and corruption. Scarce public funds are siphoned off for private ends; jobs are filled and contracts awarded on the basis of personal connections rather than merit.

The prevailing, top-down style of government is an arid climate for effective grass-roots development. Policies and programmes are all too often framed without consulting the people, and so are much less likely to command the popular support that can help them to succeed. Press freedom is often restricted, and non-governmental organizations weak, so leaders are starved of feedback on the impact of their actions at grass-roots level. Civil servants anxious for promotion are not likely to blow the whistle if things are going wrong in their area. There is a vacuum of accurate information, making it far easier to persist with misguided policies, and mistakes take much longer to correct.

A hostile policy environment can cripple individual projects just as surely as drought withers plants. Urban bias has over-shadowed all development efforts in rural Africa. Projects aimed at increasing food or cash crop production regularly fail because the prices paid to farmers are too low to give them any incentive to boost their output. In one World Bank project in West Africa farmers who were growing millet, sorghum and groundnuts were encouraged to shift over to cotton and maize, which studies had suggested would produce a higher return. But the prices of cotton and maize remained low: as soon as the project was over, farmers went back to their old crops. The vast Rahad irrigation project in the Sudan compelled farmers to grow cotton, but cotton prices were low due to low world market prices and an overvalued Sudanese currency. Two years into the project, three out of five participating farmers were making a loss.

Urban bias makes it difficult to recruit and keep good staff in rural areas, usually devoid of all mod. cons. from electricity and running water to shops and cinemas. Rural areas are still openly described by civil servants with the abusive colonial term 'bush' (*la brousse* in former French colonies). A spell of rural service is regarded almost as penal servitude or exile, to be completed as quickly as possible. Staff morale and commitment are low, and turnover is rapid.

Western bias has been just as damaging, producing a sort of blinkering, a failure to understand the very distinctive characteristics of African environments and societies. Adequate anthropological studies or the effective participation of farmers could have remedied this ignorance, but both were neglected. Misreading of social, economic and cultural factors, the World Bank reports, was a factor in virtually every project where new technology was rejected. The wholesale introduction of unfamiliar technology, or technology that required major changes in existing systems, was a common cause of failure.

THE PITFALLS OF AID

Far from alleviating these problems, foreign aid in Africa has often aggravated them.

Aid enables countries to make imports which they could not otherwise afford. It may have helped governments to maintain overvalued currencies, and reinforced Africa's dependence on imports. Aid donors have been generally reluctant to shoulder the *local* costs of projects, and have tended to boost their *import* content – whether to justify the commitment of so much foreign exchange, or to provide export markets for industries and contractors back home. This has created projects that were dependent on imports and therefore highly vulnerable in financial crises. Aid projects typically last only three to five years and donors do not usually accept responsibility for on-going costs after that period. Once projects are officially over and the aid funds dry up, governments who welcomed the chance of a subsidized programme (and the much needed foreign exchange and free jeeps or cars that came with it) are often unable to take on the full costs, and projects falter or collapse.

Official aid is supplied direct to governments, and bolsters regimes even when they are unpopular. Since aid covers the cost of many items that would have to be imported anyway, the foreign exchange saved can be used to import armaments, for repression or aggression as well as defence. Aid to government tends to reinforce government authority over against popular organizations and the private sector. It provides additional government jobs, many of them in urban areas. Unless strenuous efforts are made to avoid it, aid can strengthen centralization, sustain unpopular governments, inflate overpaid bureaucracies, and hasten the pace of urbanization.

Donors like the World Bank often complain of managerial weaknesses and lack of government commitment to projects. Yet many aspects of aid donors' own behaviour add to these problems. Donors are many: the World Bank and its component institutions; the United Nations Development Program and the UN's specialized agencies; the Eastern Bloc; various Arab and OPEC funds; plus the bilateral aid of dozens of donor countries with several voluntary aid agencies each. In the early 1980s, for example, Malawi was dealing with 50 separate donors for 188 projects; Kenya had to liaise with 60 donors on 600 projects; while in Zambia no fewer than 69 donors were involved in 614 projects. A huge share of the time of senior government personnel is taken up with dealing with all the donors, preparing accounts and reports, ferrying round visiting missions and media people (including myself). Each donor has its own policies, its own priorities, its own approaches. For example, no fewer than 11 donors were involved in drinking water supply in Tanzania, every one with its own separate procurement and supply channels and maintenance systems. As Robert Berg of the Overseas Development Council puts it: 'What makes good sense in each individual project creates cumulative chaos.' 'The proliferation of projects,' the World Bank comments, 'may actually have undermined the development effort of individual countries.'

Too many cooks spoil the broth, and to complicate matters

further, each cook switches the recipe from time to time. Fashions
and trends sweep through bilateral and multilateral programmes
as Western governments and key personnel change, and as past
failures lead to the search for new panaceas. In the first two
and a half decades of African independence, the development
consensus shifted from high-tech to low-tech to mixed-tech, from
project aid to aid designed to influence policy, from enthusiasm
for socialist or populist approaches to pushes for privatization,
from hysteria about population growth to cuts in population aid.
These policy shifts occurred quite independently of changes in
African government policies. It is not surprising if African
governments play along with each new set of strings attached to
aid, without really accepting the new melody. Hence the frequent
lack of real government commitment to projects and programmes
inspired by donors. The proliferation of donors, the inconsistency
over time, the lack of co-ordination, all add to overloading of
senior staff. They produce confusion of goals, and chaos in
procurement and maintenance.

Faced with shortages of qualified staff, donors have often been
tempted to secure special arrangements for their own pet projects,
rather than working to strengthen overall national capacities.
They have relied excessively on expatriates, whose excessive
salaries and expenses absorb far too high a proportion of costs.
Even with the best will in the world, most expatriates cannot
hope to gain a deep understanding of local problems. There has
been a temptation to attract good local staff by offering salaries
above national civil service levels. This has merely poached from
the limited pool of talent, shifting the shortages from the project
to other government departments, and stirring up problems of
morale and rapid turnover in the civil service. Area development
projects, such as those favoured by the World Bank in the 1970s,
sometimes created separate institutions and back-up agencies,
which were hard to absorb in national programmes when the
project was over.

If projects are to contribute to Africa's long-term development,
they must be sustainable, but aid procedures often neglect this
factor. The time beyond the life-span of the aid project, when
the host government will have to shoulder the on-going costs, is

ignored in the rush for fast, visible results. The handover of responsibility can be highly disruptive: overworked government agencies must now take on the tasks of well-funded project management. Government budgets have to take on the long-term costs of facilities set up under aid projects – and they may not be able to.

The combined impact of all these obstacles and errors has been devastating in Africa, especially in the central areas of agriculture and environment.

1. Not enough has been spent.
2. Much of what has been spent was inefficiently spent, its effectiveness massively reduced by management and staff problems, breakdowns and delays, import and budget cuts, unattractive producer prices and tenure arrangements.
3. Many of the things that were done were the wrong things to do, irrelevant to the central problems, or indeed liable to worsen them.
4. The wrong policies were pursued.
5. Foreign aid was inadequate, and what aid was given often helped to aggravate the weaknesses of government programmes.
6. The African peasant was virtually abandoned. The intensification that government action could have brought about did not occur. And so the destructive logic of the underlying situation, the impact of expanding populations using no-input farming, on fragile soils and in an unpredictable climate, has continued and accelerated.

PART TWO
The Response

Wherever dangers mount,
the powers that save grow, too

Hölderline, *Patmos*

4 Tapping the potential

The outlook over Africa is gloomy.

The roots of Africa's problems lie in the tangled relationship between agriculture, environment, and population growth. It is precisely these crucial areas that have seen the worst record of failure.

And yet there are breaks in the storm clouds: individual projects, sometimes countrywide programmes, that have succeeded against a background of general failure – ventures that have won through against all the multiple odds. These efforts are of crucial importance to Africa. They have clearly found some way of breaking through the impasse. If we could read the lessons of their success, we might be able to piece together the formula. That is what the rest of this book sets out to do.

How do we recognize success? There is no safe consensus on the criteria – all I can do here is spell out the checklist I used for selection. Projects had to have been going for at least two years, and to have achieved their stated goals, or overfulfilled their targets. They had to offer attractive economic rates of return, improve the long-term productivity of land or meet other basic needs, and reduce the impact of absolute poverty. They had to be ventures that enjoyed a large measure of popular support. In some case the popularity was indicated by strong public demand for expansion, with demand regularly outstripping and stimulating supply. In other cases the people themselves had taken the ball and run with it, passing an innovation spontaneously from neighbour to neighbour, or traders had spotted a mushrooming market and supplied it themselves. Projects of this kind were, like genetic material, self-replicating.

I looked for projects that were sustainable over the long term. In the African context, by and large that meant projects that were to a considerable extent self-sustaining, that could survive in an atmosphere of budgetary restraint or import restrictions. It meant projects that conserved or enhanced natural resources. If they were to have lessons for the rest of Africa, they had to be replicable in their essential features (though obviously not in detail) over a wide range of circumstances.

I have included a few exceptions that do not meet – or do not yet meet – all my criteria. These are mainly recent research or experimental efforts that look well placed to succeed and spread, because they embody the qualities of the longer-standing success stories. There are a small number of projects that are not low-cost, but not high-cost either. These are included for their importance, and on the grounds that some African countries now, and more in future, may be able to afford them.

Many of the projects we will look at are among the most outstanding in Africa. Of course they are not exhaustive: there are other ventures, in other countries, by other agencies, that are doing just as well, and no one need feel offended if their project is not included. The projects I have chosen are broadly representative. They cover every major sector in the problem areas we are concerned with. They range across the major habitats, from humid to semi-arid and mountain, though with a strong bias towards the problem environments. They include every scale, from village level to nationwide. There are African government projects and African non-governmental projects; bilateral, multilateral and voluntary aided projects. And projects in countries ranging from unashamedly capitalist Kenya and Nigeria, through the more mixed economies of Zimbabwe and Niger, to socialist Burkina Faso and communist Ethiopia. Despite the wide differences of background, we shall find many common threads in the successful projects.

In the chapters that follow we will have a double aim in view. One is to look at the 'what' of Africa's future, to explore the concrete technologies and approaches that seem most likely to work in each sector, and in conjunction, in Africa's special social and environmental conditions. The other is to examine the 'how', to try to discover more general features that seem to be essential ingredients of success in any sort of project in Africa. The composite picture, in my view, adds up to the best possible scenario for Africa's future over the next decade or two, and to a strategy that can, at one and the same time, boost food production, conserve the environment, and reduce absolute poverty. Whether it will be realized depends on all our efforts, North and South, but especially on those of African governments and of the African people themselves.

THE MASTER FARMERS

Perhaps the most convincing evidence that the people are ready for appropriate, well-thought-out change is the resourcefulness and energy of Africa's farmers. There is here a tremendous untapped potential for rapid innovation.

The conventional colonial view had it that the African farmer was a hopelessly inefficient stick-in-the-mud, unwilling to give up traditional methods that had persisted for centuries. Nothing could be further from the truth. Africa's peasants, in my experience, are among the most inventive and adaptable in the world. They have to be, because they are dealing with the most varied and unpredictable environment in the world. They are always open to new varieties, even to new crops, that can make the best of their limited resources. Long before the Europeans moved in, Africans had adopted and spread exotic plants: cassava; maize; bananas; sugar; tobacco. British anthropologist Paul Richards has shown that traditional farmers, untouched by conventional development projects, select and breed their own improved varieties, carefully adapted to the needs of their location. One survey of 98 farm households in Sierra Leone found no less than 59 distinct varieties of rice in use. Each household was planting between four and eight separate varieties, suited to the varying soil and water situations on their holdings. A smaller sample of 30 households had tested or adopted an average of 2.5 new varieties each over the previous ten years.

In no other continent is there such a diversity of farming methods and systems. Often each household will be using two or three systems at once – on top of the simplified pattern of shifting cultivation we examined in Chapter 2, they may be exploiting the limited manure they possess on the compound close to their house, and using shallow groundwater in the nearest valley bottom. They have developed dozens of ways of moulding the soil, planting sometimes in ridges and mounds in wet areas to improve drainage, sometimes in furrows and hollows in dry areas, to collect scarce water.

Patterns of intercropping are even more diverse. In northern Nigeria as many as 156 separate crop combinations have been observed. The Konso people of South Eastern Ethiopia regularly cultivate 49 different plants, shrubs and trees. In a single small Konso field of 0.2 hectares, 24 different species were counted, including sorghum, maize, barley, millet, amaranth, tobacco, coffee, cotton, pigeon pea, kidney beans, lima, castor and hyacinth beans, taro, yam, cassava, safflower, linseed, red pepper, tree gourd, and moringa, a tree whose leaves are boiled as vegetables.

In many areas where population density is very high, African farmers have developed their own distinctive forms of intensive agriculture. These too are of a bewildering diversity.

In the highlands of East Africa, where trypanosomiasis is not a problem, mixed farming systems have developed, closely integrating cattle and crop-farming. In some very densely populated areas like the slopes of Mount Kilimanjaro and Mount Kenya, or Ukara Island in Lake Victoria, cattle are no longer left to graze freely. They are stall fed on plant refuse and cut fodder, and their dung and bedding is used as manure. Because the cattle waste less energy moving around, they produce a lot more meat and milk and no less environmental damage.

Irrigation is practised, though on a very small scale. Valley-bottom lands, with richer soils washed down from surrounding slopes, are widely used in the dry season to grow vegetables, hand-watered from shallow wells. In Sierra Leone, rice is grown in tidally flooded mangrove swamps. In Tanzania the flood plain of the Rufiji river is cultivated: crop roots grow down as the water table sinks gradually. There are even a few isolated examples of more sophisticated approaches, not unlike the village irrigation systems of Indonesia. The Sambaa and the Para of northern Tanzania build long channels out of stone.

Organic recycling – use of crop residues and manure – is found to some degree almost everywhere. It is usually confined to the home garden area around the compound, but some areas have more advanced systems. Shona farmers in Zimbabwe put their maize stalks on the floor of the kraals where they keep their cattle at night to collect dung and urine. After a few months a rich grey compost forms which is used on vegetable gardens. The Bukabo of Tanzania used to green manure their impoverished laterite soils by carrying grasses and reeds from swamps and marshes and using them as mulch. The Matengo, who live in a mountainous area of southern Tanzania, combine organic recycling with soil conservation. They dig out pits three feet across. The turves removed from the surface are then stacked in ridges into which crops are planted. The pits and ridges collect all soil washed away by the rains. And all organic wastes are thrown into the pits where they compost naturally. The location of pits and ridges are regularly exchanged. The system creates a rich and fertile soil, loses very little water in run-off, and virtually eliminates soil erosion despite the hilly terrain.

The fourth African path to intensification is agroforestry: the combination of trees and crops, sometimes with livestock. In the plain around the town of Debre Zeit, in Shoa province, Ethiopia, the fields are scattered with *Acacia albida* trees. They are not planted, but fostered by farmers wherever they sprout up. The tree keeps its leaves in the dry season and serves as browse for livestock when little other fodder is available. At the end of the dry season, the leanest time of the year, it drops protein-rich pods – up to 5 tonnes per hectare – which can be used for fodder. When the rainy season arrives, the tree obligingly sheds its leaves, so it does not compete with crops for light. Soil around the trees is enriched by the tree's nitrogen-fixing root nodules, by leaf fall, and by the droppings of livestock that shelter from the hot sun. Yields of crops around the trees are double those further away.

In south-eastern Nigeria, one of the most densely settled parts of Africa, forest farmers have developed sophisticated systems of tree and crop farming that mimic the multi-storeyed vegetation of the rainforest that surrounds them. High overhead grow tall coconut and oil palms, with medium-sized trees like breadfruit, raffia, and pear immediately below. Then comes a layer of shorter trees like kolanut, mango, orange and lime, with pawpaw, bananas and plantains below them. Yam vines grow up stakes

The rainforest is layered in levels that break the force of rain and keep soil temperatures lower, while leaf littler maintains soil structure and fertility

to 3–6 metres, while cassava, cocoyam and pepper bushes reach shoulder height. In the scattered clearings grows maize, intermixed with ground crops like peanuts, melons, and vegetables. The system looks like a jungle, but everything is carefully placed to exploit small differences in soil and water availability, light and shade. Labour needs are spread more evenly, some food is available all the year round, and the soil is rarely exposed to rain. The system is highly productive – tree density is as high as on commercial plantations. Plants are closely spaced to increase soil cover. All the trees are regularly pruned and trimmed. Lopped branches are used for yam stakes, building poles and firewood. Twigs and leaves are fed to goats or used as mulch. Bananas, which thrive on organic material, are planted in the old sites of household latrines.

A few tribes have combined irrigation, agroforestry, organic recycling and soil conservation in systems of extraordinary sophistication. The Dogon, who live along the escarpment of southern central Mali, planted *acacia* trees 40 to 50 to the hectare to reduce erosion and increase soil organic matter. They built terraces to catch the rain to retain soil, and enriched the soil with earth collected from fertile river valleys. They dug large pits and lined them with boulders and stones to create small reservoirs. They composted ash, dung, millet dust, stalks, and household refuse, into a high-quality organic fertilizer. The Dogon are experts at storage, building enormous granaries in which grain can be kept for up to seven years as a protection against famine.

Perhaps the most highly developed traditional system was evolved by the Chagga, who farm the southern slopes of Mount

Multi-layer farming reproduces the rainforest's layers with intermingled crops that have complementary needs for light, water and nutrients

Kilimanjaro. For at least a century the Chagga have made careful use of melting snows, tapping the waters of high mountain streams into networks of skilfully aligned irrigation channels. Certain clans specialize in the design of furrows, while all Chagga share responsibility for their maintenance. From the nineteenth century the Chagga have been stall-feeding their cattle and using the manure on their plots. They also developed their own system of multi-layer farming, growing intercropped yams, sweet potatoes, maize, beans and peas among coffee bushes, with banana palms providing shade.

If these intensive systems had been applied on a wide scale, Africa would not now be facing ecological crisis. Unfortunately they are usually very localized, and have not spread far beyond their point of origin. One reason may be that while seeds or new crops can easily be traded or transported, farming systems are complex ways of managing crops and soils, and have to be *taught*. Their spread would be slowed to a trickle at every tribal or language boundary, every desert or watershed separating largely self-sufficient groups. For example, the *acacia albida* of the Debre Zeit region disappear suddenly as you leave the surrounding catchment. I asked one seventy-year-old farmer why people in the next valley did not use the tree. He told me he was not aware that they did not use it. He had never travelled outside his home area.

The main reason why these systems have not spread far is that they arose in special circumstances, in areas where population density was high, often because local people were confined to a small space for security reasons, or because they were in a better-watered hilly area surrounded on all sides by dry zones. They intensified, as most pre-modern farmers in Europe or Asia did, quite simply because they had no choice. Their population densities had reached the point where they had to go on farming the same piece of land more or less permanently. They had to put back most of the nutrients the crops took out, and they had to conserve soil and water.

Eventually, population density will compel all African farmers to intensify, but if we wait for this process to take its course, a great deal of Africa's land will have been damaged irreparably, and holdings will be so small that they will still afford no more than a subsistence.

The only way Africa can escape this gloomy future within the next decade or two is if *her farmers intensify before population density compels them to do so*. Two prerequisites are needed if this is to happen. The first is *incentives*. The carrot of increased incomes, of manufactured goods and devices that save on back-breaking labour, can replace the stick of overcrowding. But this will only happen if governments and international markets can offer fair prices for the farmer's produce. The second is a more rapid diffusion of intensive and soil-conserving techniques by way of nationwide extension systems in which extension workers help farmers to combine the best of traditional and improved methods.

African smallholders have a lot to learn from the best practice of their fellows. And they also have a lot to teach. They have proved themselves masters in managing natural diversity, in adapting to a wide variety of ecological niches, and in coping with a higher level of risk than is faced in any other region of the world. They are masters of economy in the use of capital and labour. They are fast learners, and inveterate experimenters.

Left to themselves, they will probably adapt fast enough at least to survive, barring further droughts. Given the right incentives, they will adapt more speedily and more intensive techniques will spread spontaneously. But given modest extra resources and guidance, spread through nationwide networks of advice and supply, they are capable of rapid advances that could totally transform Africa's food outlook within a decade.

5   Boosting food production

The cutting blade of Africa's crisis is the steady decline in agricultural production per person, especially of food. The threat is double-edged: on the one side it reduces the incomes of the rural majority, and their access to food. On the other, it reduces the availability of foreign exchange to finance industrialization, and cuts local demand for manufactured products. Africa's agricultural predicament threatens not only to perpetuate the poverty of her subsistence farmers, but to forestall her prospects of overall development.

Let us recall the basic trends set out in Chapter 1 (p. 20). Between 1965 and 1982, African food production per person fell by 12 per cent while in every other developing region it rose. The decline occurred in no less than 33 out of 38 sub-Saharan countries. It is time to examine the causes in more detail.

Neither drought nor declining rainfall in the Sahel are among the fundamental factors. The decline affected all regions. It was most pronounced in Southern Africa, where per capita food production fell by an average of 19 per cent between 1969–71 and 1981–2. Anti-colonial wars and civil wars in Angola, Mozambique and Zimbabwe aggravated the situation here, along with South Africa's efforts to destabilize her black African neighbours. The fall in the Sahel was 13 per cent. Declining rainfall levels must have played some part in this region, perhaps accounting for around 5 per cent of the drop. But more general factors were at play, for the two geographically most favoured regions – humid Central Africa, and cooler East Africa with its superior soils – also registered drops, averaging 8 per cent in each case. In humid West Africa, the fall averaged 11.5 per cent – weighted down by a massive drop of 36 per cent in Ghana, where economic mismanagement was a major factor. Without Ghana, the regional average here, too, was 8 per cent.

Clearly, continent-wide factors were at work. It is fashionable to cite the spread of cash crops as one of these. Cash crops, so the argument goes, take up the best land and the major inputs. Food crops are squeezed more and more onto marginal land. This theory may apply in a few cases, such as large-scale irrigation schemes in the Sahel or the few countries like Kenya where land ownership is very unequal. But as a general explanation it does not hold water. In most cases, cash crops have not prospered at the expense of food crops: often, they have performed even worse. The total area under cash crops has in fact steadily decreased in Africa as a whole. Between 1969–71 and 1979–81 the total area under the nine main cash crops[1] *fell* by 9 per cent, while the total arable area *expanded* by 6 per cent.

The biggest drops in area were for the two crops most frequently accused of responsibility for the Sahel disaster: the area under groundnuts declined by 18 per cent, and that under cotton by 26 per cent. In the five Sahelian countries of Senegal, Mali, Burkina Faso, Niger and Chad, the proportion of total arable land taken up by these two crops fell from 17 per cent in 1969–71 to 12 per cent in 1983. Yields of sesame, groundnuts, cocoa, coffee and sisal dropped faster than those of any food crop in the ten years to 1979–81, as did the production per person of these five plus cotton. These trends are not surprising, as the prices of most of these commodities fell, often sharply.

To understand the real causes for the slow decline of food production per person in Africa, we have to break the problem into its constituent parts. The rate of growth of *per capita* food production is made up of the rate of growth of *total* food production, minus the *population* growth rate. The first point to note is that total food production in Africa actually rose, at a rate of 2.1 per cent a year through the 1970s. This was no mean achievement – it was, surprisingly enough, faster than the 1.8 per cent a year achieved by developed countries. What dragged the food production *per person* down in Africa was the phenomenal

[1] Groundnuts, cotton, cocoa, coffee, sesame, sisal, sugar, tobacco and tea, in order of area cultivated.

rate of population growth, averaging 3 per cent a year in the 1970s. Rapid population growth, therefore, is an important factor in the equation.

Another way of looking at the same problem is to ask why food production in Africa failed to keep pace with population, while it more than kept up in the other developing regions. Increase in food production is due to two components: expansion of the cultivated area, and increases in yield from a given piece of land, usually by growth in irrigation, use of fertilizers and pesticides, and improved seeds. In the Far East, three quarters of the growth in cereal production through the 1970s was due to improved yields, and only one quarter to land expansion. In Latin America, despite its vast land reserves, yield improvements accounted for two thirds of the production increase, and expansion for one third.

In Africa, by contrast, yield improvements accounted for only 5 per cent of the increase in cereal production. Yields of rice, millet and sorghum rose by only 0.6–1.5 per cent over the whole decade – a difference that could easily be due to climatic vagaries or statistical quirks. Maize and cassava yields both fell by 5 per cent.

Land expansion, therefore, accounted for almost all the increase in production in Africa. But the cultivated area in Africa has grown much more slowly than the population. Between 1975 and 1983 the farmed area expanded by a mere 0.59 per cent a year. Africa still cultivates only about a quarter of the land that *could* be cultivated, but her reserves are mainly under tropical rainforest and concentrated in the well-watered lands of Central and Southern-Central Africa. Here the progress of opening up new land is slowed by the inaccessibility of virgin areas, and the sparse populations. Human and livestock diseases and lack of all basic services are major deterrents. In the more densely populated mountainous and semi-arid areas, land reserves are dwindling fast. Farmers are pressing up against the limits of cultivation, and sometimes beyond, farming more and more marginal drylands and steep slopes. Nigeria, Burundi, Burkina Faso, Mali and Sierra Leone are all using more than three quarters of their cultivable land, while Rwanda, Togo and Niger are farming more than nine tenths.

The heart of the problem, then, is the fact that food crop yields have stagnated or declined in Africa. The basic reason, as we have seen, is that most African farmers are still practising the no-input methods of shifting agriculture, while fallow periods are being cut back to the point where nature is no longer given enough time to restore soil fertility. A gradual increase in the use of inputs like fertilizers should compensate for the dwindling fallow periods, but this has not happened fast enough in Africa; indeed, in many parts it has not happened at all. Africa uses less fertilizer than any other region – less than 10 kg per hectare in 1983, against 37 kg in Latin America, 41 kg in Asia and 46 kg in the Near East. Fertilizers can be organic as well as chemical, but in many areas crop residue and animal manure are increasingly burned as fuel as tree stocks dwindle. In Niger, millet stalks that could protect and enrich the soil are used for fences or roofs, for cooking or for livestock fodder. In highland Ethiopia, mothers and children scour the fields at dusk for dried animal droppings, mix them with straw, and mould them into fat pancakes, for burning. More than half of African countries use less than 5 kg of fertilizer per hectare, and much of that goes on cash crops – in Tanzania, for example, for every kilo of fertilizer that went on each hectare of maize in the mid-1970s, 21 kg went on sugar, 35 kg on tea, and 46 kg on tobacco. In the Sahelian countries, 59 kg of fertilizer went on every hectare of export crops in 1979–81, but only 2 kg on food crops. Food crops fared better in West Africa and Southern Africa, but even here cash crops got three to six times more fertilizer per hectare.

Most regions of Africa have still not passed the critical population density at which they would be forced to intensify. In the modern world, the carrot of manufactured goods can push farmers to intensify before the stick of overcrowding forces them to. This depends, to some extent, on governments forcing the pace by providing inputs, advice, credit and incentives. But most African governments have grossly neglected agriculture. In nine out of ten African countries studied by the FAO, agriculture got less than 15 per cent of government spending in 1978–82. The average was 10 per cent – though agriculture accounted for an average of almost 40 per cent of the gross domestic product in the ten countries. A great deal of spending on agriculture is wasted or used inefficiently – 38 per cent of it goes on administration, for example, compared with 11 per cent in Latin America and only 3 per cent in Asia. And much of it is very badly spent.

We have already seen the appalling record of agricultural projects in Africa (p. 47). Projects were designed in blithe oblivion of Africa's social, economic and environmental realities. Until recently little or no effort was put into studying what African farmers did, and why. It was assumed that most of what they did was primitive, inefficient, even destructive. Western practices were pushed but many proved worse than useless in Africa's environment. Heavy machinery was used, destroying the fragile structure of African soils. The valuable practice of intercropping was discouraged in favour of sole cropping. Chemical fertilizers were used which acidified the soil in humid areas, and in dry areas often yielded less in extra production than they cost to buy.

The realities of life for African peasants were often ignored. Projects were pushed that expected farmers to invest cash that most of them did not have – or to take on risks that could jeopardize their prosperity and even their existence. As World Bank forestry expert Jean Gorse has pointed out, project planners usually neglected the fundamental significance of rainfall variability and risk avoidance. They introduced technologies that might do better than traditional approaches in years when the rains were good – but that might do disastrously *worse* in a drought year.

Sometimes new crops, new varieties, new technologies were introduced individually without reference to the complex farming systems they would become part of. Worse still, whole packages of exotic species and foreign technologies were launched on unfortunate farmers, based on results from other continents, or from isolated research stations. The packages were unfamiliar, or required radical upheavals in the farmers' whole system. Not surprisingly they were restricted to small areas, or misapplied.

Too often, projects have focused on big, high-profile, high-cost ventures – highly mechanized state farms and plantations, vast irrigation schemes – which floundered, ignoring the fact that small farms account for the overwhelming majority of food production in Africa. Projects have assumed continuing levels of supervision, management, supplies and extension, which governments with their shaky budgets simply could not sustain. They have relied too much on imports of fertilizers, machinery, and spare parts, which African governments, with their crippling shortage of foreign exchange, often had to cut back.

One of the major sources of failure has been the economic environment. Even where inputs were provided, farmers often had no real incentive to buy them or use them, because the prices they could get for their produce were not attractive. As we have seen, the prices farmers received for cash crops were kept low by governments siphoning off a high proportion of the market value, already depressed by declining world prices for agricultural commodities and by overvalued exchange rates. The prices of their food crops were kept low by low official procurement prices, by cheap imports and food aid, and by urban consumers' tastes shifting towards cereals that were hardly grown in Africa. Farmers had little incentive to produce more food than they needed for their own subsistence. And they had little or no spare cash to invest in inputs like fertilizers or pesticides. Even when they did have cash, low prices meant that it was uneconomical to invest in intensification. Low farm prices perpetuate no-input farming and the mining of the soil. They are a major cause of Africa's environmental crisis.

LESSONS FROM THE LEADERS

The landscape has not been uniformly bleak. Five countries did manage to increase their per capita food production against the general trend of decline. These were Swaziland, Rwanda, Ivory Coast, Malawi, and Sudan.

In Sudan and Ivory Coast, the *whole* of the production increase was due to an increase in the cultivated area. Yields of both cereals and root crops declined. Progress under these conditions is not sustainable: it will halt and reverse as soon as the limits in land expansion are reached.

Swaziland, Malawi and Rwanda are genuine success stories with important, and surprisingly diverse, lessons for other African countries. Swaziland achieved the highest increase in food production per person – 16 per cent, between 1969–71 and

1981–2. The *whole* of this increase was due to yield improvements. Cereal yields, for example, rose by 50 per cent in the decade to 1979–81. The advance was largely due to high inputs – around 1980, for example, Swaziland used 120 kilos of fertilizer per hectare, which is higher than the United States. Some 15 per cent of Swaziland's cultivated area is irrigated.

Malawi boosted per capita food production by only 3 per cent over the period. Yet in many ways her success is even more striking. Three quarters of her cereal production increase was due to yield improvements and only one quarter to expansion in the land area. Malawi managed to improve cereal yields by 18 per cent between 1969–71 and 1979–81. In 1970 she was a net cereal importer to the tune of $7.8 million a year; by 1983 she had become an exporter, earning a net $5.5 million from her cereal trade. And yet Malawi uses little more chemical fertilizer than the African average – 14 kilos per hectare in 1982 – and only 0.3 per cent of her farmed area is irrigated. The success here is primarily due to trade and exchange policies that have favoured food producers and exporters. Prices paid to farmers for maize have been one third or more *above* world market levels, and Malawi has had one of the most open markets and one of the least overvalued currencies in Africa. Indeed, a World Bank study of price distortions due to protectionism, exchange and interest rate policies in 31 developing countries ranked Malawi top by a clear head, beating even high performers like South Korea and Malaysia. Malawi has also given a very high priority to agriculture, which got 15 per cent of the national budget in 1982–3 – a nose ahead of education, and double the share of defence.

Rwanda's story is remarkable in a very different way. Per capita food production rose by 8 per cent over the period 1969–71 to 1981–2. Yield improvements were modest for cereals, only 3 per cent in the decade to 1979-81, but Rwanda's staple foods are root crops, and their yields improved by 13 per cent. Indeed, the yield of sweet potatoes, the principal root crop, rose by 35 per cent, while yields of coffee, the country's main export, improved by 49 per cent. And yet Rwanda used virtually no chemical fertilizer or irrigation. Her secret has been an intensification of farming on purely organic lines with a strong emphasis on conservation. Population density is so high that fallowing has virtually disappeared. Soil fertility is maintained by heavy use of mulches on coffee groves, and of crop residues dug into cultivated fields. Ridges along the contours prevent soil erosion on hilly land, while the emphasis on tree and shrub crops – bananas, coffee and cassava – protects the soil against erosive rainstorms. It is doubtful whether Rwanda's agriculture and economy can take off relying solely on these methods. But they have allowed her farming sector to keep pace with population growth while conserving the soil. When she *does* manage to introduce chemical fertilizers and improved varieties on a wide scale, organic and soil-conserving methods will provide a far more solid base for further progress.

ZIMBABWE'S MAIZE MIRACLE

The most dramatic transformation in African agriculture has occurred among Zimbabwe's black farmers, since independence in 1980. Within the space of just five years, many of Zimbabwe's peasants have moved from subsistence to dynamic commercial farming, from yields well below the low continental average to yields in some cases ahead of large-scale white commercial farmers.

Travellers in Zimbabwe in the nineteenth century wrote accounts of impressive prosperity, of valleys converted into the most fruitful of gardens, of huge underground granaries. But from the 1890s white settlers began to seize vast tracts of the best land. Africans were converted into tenants on their own ancestral soils: many fled into the native reserves. By 1930 white farmers controlled half the country. The division, formalized in the 1969 Land Tenure Act, gave 16 million hectares to 6,700 European farmers. More than half of that was good land suitable for cropping. The same area of land was allocated to one hundred times as many black farmers, but three quarters of their lands were poor and marginal, more suited to pasture.

This division of land still persists largely unchanged in 1986. The transition on the ground between the European areas and the communal lands is often as distinct as a frontier on a map, especially in the dry season. On the road east from the capital

Harare to Murewa, the white commercial farms, with rich red soils, and dense forests surrounding healthy stands of wheat, irrigated from sprinklers like arrays of Baroque fountains, suddenly give way to dry, sandy soils with mud huts scattered among rocky outcrops, in many places overcrowded with people and livestock, and with only scattered trees.

Not just ownership, but the whole colonial agricultural system was skewed to suit the whites. In 1972, for example, 87 per cent of agricultural credit went to the tiny minority of whites. By 1979, there was one agricultural extension worker for every 70 European farmers, against one for every 1,000 communal farmers. The whites got preferential prices for their maize. Not surprisingly, the Europeans made far faster progress than the blacks: their maize yields rose more than twice as fast, their sorghum yields nearly six times as fast. By 1979, the white farmers were getting maize yields of 5–6 tonnes a hectare – on a par with Canada or France. The neglected communal farmers got a meagre 0.6–0.7 tonnes a hectare – a good way below the all-African average of 0.95 tonnes.

White rule came to an end in 1980, after years of armed struggle, and a new black government under Robert Mugabe took over. There were no recriminations, and despite the historical justification, no wholesale expropriations of white farms. But there was a total transformation of the government's agricultural policy, a re-orientation towards the black farmer who had been discriminated against for so long.

The results were astonishing. In 1978, Zimbabwe's black farmers produced only 514,000 tonnes of maize. Of this, they sold 64,000 tonnes – a mere 7 per cent of the marketed total. In 1981, after only one year of independence, they more than doubled their production to 1,054,000 tonnes. Over 360,000 tonnes of this were put on the market, and made up 18 per cent of total deliveries to the grain marketing board (GMB). Drought hit the maize crop over the next two years. But in 1985 Zimbabwe's black farmers produced an astonishing harvest of 1,780,000 tonnes. Of this, they sold no less than 939,000 tonnes – 48 per cent of the GMB's intake. Yields are difficult to estimate, but probably averaged around 1.3 tonnes a hectare – an advance of more than 80 per cent on the 1979 level. Many farmers did far better. Most of those I met had got between 2.5 and 4.3 tonnes a hectare in 1985.

Jerry Guyo, chairman of the farmers' group in Chidembo, topped them all. The village is far off the beaten track, up and down hill along a dirt road rutted with deep gullies. Guyo has a restrained, controlled energy and a profound knowledge of farming boosted by two years of spare-time training for the coveted Master Farmer's certificate. When I visited him, his compound yard was neatly stacked with no less than 184 200-lb bags of maize, grown from just 4 acres (1.6 hectares) – an astonishing yield of over 10 tonnes per hectare, right up with the world's top maize growers. Before independence, he was happy with yields one sixth as high.

On another three acres he grew 28 bales of cotton, worth 3,000 Zimbabwe dollars (around US $2,000). He was planning to sell 170 bags of maize, which would bring another Z $3,060. All that was not achieved without high inputs: a massive 500 kilos of fertilizer per hectare (equal to Belgian levels), high-yielding hybrid seeds, and pesticides for the cotton. But his total outlay on inputs was only $905 – most of that covered by a loan from the government-backed Agricultural Finance Corporation (AFC). His net cash income was over $5,000 (around US $3,300) – at least ten times his pre-independence income, representing a return on his cash investment of a handsome 400 per cent.

Jerry Guyo has not squandered his new-found prosperity. He bought his wife some new pots and cutlery, and a smart blue overall for himself. But he still goes barefoot. Most of his profits have been re-invested. He bought a harrow, and a maize sheller to cope with his huge harvests (he hires it out to neighbours). In 1986 he will be able to pay cash for all his inputs, and will not need a loan from the AFC.

PULLING OUT ALL THE STOPS

Many elements have to come together for the alchemy of rapid growth in yields: Zimbabwe's programme has assembled them all. The first is inputs, especially fertilizer and high-yielding seeds. Africa's first high-yielding hybrid maize, SR52, was bred in what was then Rhodesia way back in 1949; since then the

country has produced its own high-yielding seed. It was the first
country in the world after the USA to do so. During the time of
illegal independence, sanctions forced Rhodesia to produce most
of her own manufactured requirements and she began to make
her own fertilizers, using hydroelectric power instead of oil as
the feedstock. Flourishing private companies, which grew up to
sell supplies to white farmers, have now expanded their outlets
into the communal areas.

A package of inputs was formulated for the small farmers: the
pack for a half hectare of maize includes 250 kilos of fertilizer
and ten kilos of hybrid seeds – a mixture of short- and medium-
maturing varieties that will yield something even in a year of
poor rains, thus providing farmers with their essential insurance
against drought.

The major problem in pushing high inputs among African
subsistence farmers is finance. They have no surplus cash. The
answer, of course, is credit: if they can only borrow enough for
one season's inputs, they can repay the loan from the increased
harvest. Credit programmes in Africa face a number of obstacles.
Under traditional systems of land tenure farmers have no formal
title to the land which could serve as collateral. The answer
Zimbabwe has adopted is to use the eventual crop as collateral,
and deduct the owings from the payment for the crop. The
repayment record has been good, though inevitably drought
years present a problem: repayments slump, and in the following
year farmers are reluctant to add to their debt burden.

Zimbabwe farmers seeking credit must be members of farmers'
groups, which ensure collective responsibility for the loan. They
apply to the government-run Agricultural Finance Corporation,
which sends round assessors to check that applicants are *bona fide*
farmers. The loan is granted only for the purchase of the
recommended package of inputs. It is not paid to the farmer, but
goes direct to the seed and fertilizer supply companies, who
organize deliveries and make sure they arrive on time. The AFC
has increased loans to African smallholdings from a mere 4,400
in 1979–80 to 96,000 in 1985–6.

After inputs and the credit to buy them with, the next essential
is to provide farmers with sound advice on how to use the inputs
by way of agricultural extension workers. Those workers, in turn,

must be backed up by a research effort which is attuned to the small farmer's needs and limitations. Zimbabwe has one of the most dynamic and effective systems of agricultural extension in Africa. A poster on the wall of the Harare headquartes summarizes the old colonial extension service's view of the native outlook: 'Lack of innovativeness, limited view of the world, hostility towards government authority, fatalism, limited aspirations.' These were, of course, natural attitudes among black farmers starved of the means for self-advancement by an oppressive racist government. The new motto is far more appropriate for African realities: 'Build on what farmers know and use what they have.'

The ratio of extension workers to communal farmers has not been greatly improved – in 1985 it stood at about one to 800 – but the service now reaches farmers much more efficiently, working primarily with farmers' groups rather than individuals. The extension worker begins by analysing the main problems of the district he or she works in, and sets objectives based on the results that the best 10 per cent of local farmers are already realizing. In the Wedza area the work is strengthened by farm extension promoters – working farmers, elected by their group, and paid a modest salary of $75 a month. Every fortnight extension workers visit each of their groups on a fixed schedule, so supervisors know exactly where they are at any time. The farmers serve as the front line of supervision, and soon raise an outcry if the extension worker does not show up. The farm extension promoter fills in between visits, answering farmers' individual problems on their land.

Availability of inputs and advice, however, will have little or no effect on surplus *output* unless farmers have an incentive to increase production beyond their immediate needs. Without incentives, they will ignore the advice and refuse to buy the inputs. Incentives, in turn, depend on attractive prices, access to markets, and availability of goods on which to spend increased earnings.

Zimbabwe has provided ample incentives. The price the farmer receives for one tonne of maize rose steeply fom $53 in 1978–9 to $120 in 1981–2 – an increase, in real terms, of 60 per cent. The National Farmers' Association, which groups the communal farmers, now has as much say in the price of maize as the politically dominant white farmers used to have in colonial days. For 1985–6, the NFA requested $180 a tonne, and got it in full. Here again, an official quoted price can only serve as an incentive if the farmer has assured access to a market where he can get that price. In Zimbabwe, grain production is bought by the official Grain Marketing Board, which has reached out to the communal farmers. The GMB has 55 major depots, with 140 smaller collection points scattered around the communal areas so that, in theory, no farmers need travel more than 20 kilometres to sell their produce. In the bumper year of 1985 these depots were bursting at the seams. One depot I visited at Murewa was piled with almost a million sacks of grain, stacked in vast heaps the size of aircraft hangars, covered with tarpaulins in case of an unexpected downpour. Four years earlier, after a good season, the depot collected only 161,000 bags.

And there was plenty for farmers to spend their money on. The town of Murewa is a flourishing centre, with clothing shops, cabinet makers, supermarkets, and a farmers' shop selling everything from fertilizer to fungicide, from anti-tick ointment to ploughshares. The new purchasing power of farmers, widely spread across the country, creates jobs in manufacturing and services.

WORKING TOGETHER

One of the biggest obstacles to progress in African agriculture is the severe labour shortage in rural areas. Families that own livestock have to divert part of their labour power to herding. Only a third of farm families in Zimbabwe own draught animals. More than half of them have at least one adult member away, working in cities, mines or on white farms. In many areas half the households have female heads, who combine the heavy chores of wood- and water-fetching with the burden of farmwork. The massive rise in primary and secondary school enrolment since 1980 has taken away a prime source of family labour. Shortage of labour and animal power can mean late ploughing, late planting, late weeding and serious yield losses.

The farmers' groups which have sprung up all over Zimbabwe play a key role in coping with the labour shortage. They share tools and draught oxen. One member herds the livestock, while another cooks for the whole group. One attends the extension worker's briefing and passes the message on to the rest, another goes to town to buy or sell for everyone.

Mavis Katumba, a 50-year-old farmer and mother of ten, has five acres of land at Marerengwa, but she grows crops on less than two. She belongs to a farmers' group of 100 people. The nearest place they can buy supplies is Murewa, 10 kilometres away, and transport is so infrequent that they used to waste whole days getting there and back. Now they pool their orders and send just one or two people. For labour purposes, they are divided into fifteen smaller groups, of four to ten people: these are self-selected, to make sure they can work well together. Three people out of eight in Mavis' work group have no oxen: 'We let them use our beasts,' she says, 'so we can raise them up. Then they will get enough money to buy their own oxen. We work much faster as a group. We plant the same day we plough.'

Before independence, Mavis used to buy hybrid seed, but she could afford to renew it only once every three years: after the first year, it cross-fertilized with local varieties and gradually lost its high-yielding characteristics. She had enough manure for only half an acre. Her yields were only 0.6 tonnes a hectare. In 1985 she used the full package of fertilizers and hybrid seeds, and insured herself by planting three maize varieties – one short-maturing, in case of poor rains, one medium and one long which would do well if the rains were good. She got 33 bags of maize – about 2.5 tonnes a hectare. 'The price is so good now,' she explains, 'you can bank the money and draw on it when you need it. I don't need loans any more.'

An extensive study in the Wedza communal area documented the rapid expansion and benefits of farmers' groups. All farmers in the area made progress. In 1981–2 some 18 per cent of farmers got loans, and 41 per cent extension advice. Two years later the proportion getting loans had risen to 38 per cent, and 73 per cent got advice. Over this same period maize yields rose by no less than 50 per cent. But farmers' group members did significantly better than non-members. Thanks to the labour-sharing arrangements, they got six times more outside labour on their fields than did non-members. They hired trucks to get their seeds and fertilizers on time. They harvested an average of 2.5

tonnes of maize per hectare, almost 50 per cent more than non-group members.

Of course, not all groups work smoothly. The Chidembo group, of which Jerry Guyo is chairman, is only just re-forming after a crisis in 1985. Those who had less land objected to labour-sharing – because they had to put in more time for farmers with larger holdings. They said they were helping to make the rich richer. The *coup de grâce* came when the former leaders misappropriated the group's bank account, and members lost an average of $30 each. Despite this shock, the advantages of group membership are so overwhelming that a new group is forming around Guyo.

Zimbabwe has not resolved all her problems. One outstanding concern is the grossly unequal land distribution: the Lancaster House agreement, under which Britain granted independence, rules out *expropriation* of white farms, but Britain has provided only a fraction of the funds needed to *buy* the white farms. So far a mere 36,000 black families have been resettled on purchased white farms. As the population density increases in the communal areas, there will be growing pressures to take over the white farms, with or without compensation, especially now that the black farmers have shown themselves fully capable of maintaining Zimbabwe's food supplies.

Other problems arose from the sudden surge in output in 1985. There simply were not enough trucks and lorries in the country to shift the harvest from the farms to the depots fast enough. Many farmers I met had been waiting for up to three months for a lorry, increasingly anxious that early short rains might spoil their crops. The marketed harvest of 2 million tonnes was more than double Zimbabwe's own needs. Some of the surplus was exported commercially, some under imaginative three-way swaps in which Western donors gave Zimbabwe wheat and took her maize to give as food aid to her southern African neighbours. But outside of Southern Africa, Zimbabwe's maize cannot compete with cheap exports from the USA. It seemed likely that Zimbabwe might be saddled with an unsaleable surplus of 700,000 tonnes. In future years, she may have to use more of her maize for cattle feed, or corn oil, and diversify into other cash crops.

The surge of progress has not yet reached all Zimbabwe's farmers. The south of the country is too dry for maize, and fertilizer use is riskier here: the research agency has yet to develop a package of inputs suitable for these areas. Nor have women benefited fully. Although they make up the majority of Zimbabwe's farmers, they still complain that their husbands decide what crops should be grown. Men give priority to crops such as maize which, traditionally, belong to them, and are less interested in crops like groundnuts, the cash from which belongs to women.

What has been achieved, however, is a rapid and spectacular shift which shows what could happen elsewhere in Africa if the right moves were made. Zimbabwe's communal farmers were perfectly typical of traditional subsistence farmers, with communal ownership of land, using low- or no-input shifting agriculture. Yet no sooner were they given the wherewithal and the incentive than they were transformed into high-input, market-minded farmers with yields two to eight times higher than before; from people with no surplus, and therefore no horizon beyond next year, to people with bank accounts and plans to invest in donkey carts, improved tools and small-scale machinery.

Of course, Zimbabwe had in-built advantages. The groundwork laid by white farmers was there to turn to the service of Africans. The country was already producing its own fertilizers and hybrid seeds, and so was not dependent on imports. It had a number of private companies supplying white farmers, and the new government sensibly used these to get supplies to the communal areas. It had a significant agricultural research programme.

But there are many key ingredients in the success of the programme that could be repeated elsewhere in Africa. The first element is a clear focus on smallholders, who make up the overwhelming majority of food producers and food consumers in almost all African countries. *If smallholders can be helped, the problems of insufficient food production, of malnutrition, and of widespread absolute poverty are attacked simultaneously.*

Zimbabwe's programme does not, at this point, involve widespread participation. The input packages and techniques are standardized. But they have been based on careful on-farm research which makes sure that they correspond to the farmer's real possibilities. The mix of short- and medium-maturing maize varieties provides a cover against bad rains. And the economics of the package are extremely attractive to participants. People do not have to invest cash up front, but get credit to be repaid out of the crop marketed. No crop surplus, no repayment. The risks, therefore, are not excessive: the farmer is not expected to put his neck on the line.

The standard maize inputs for one hectare cost Zimbabwe $226 in 1985; the typical yield increase of 2–3 tonnes per hectare was worth $360–$540, while the skilled, hard-working farmer could achieve much more. *Thus the average return on the farmer's investment was between 60–140 per cent. As we shall see, returns of this order are essential for success in Africa. People at subsistence level will not take a gamble for less.*

All African countries can begin by changing the economic environment of farmers, by improving the prices they are paid for their produce and gradually reducing the competition they face from cheap food imports.

Just as Zimbabwe has achieved astonishing results by shifting the bias of her agricultural services from large-scale white farmers to the neglected blacks, all African governments could shift resources into agriculture from less essential sectors. They can all create farmers' groups to pool resources and improve the productivity of labour. They can all set up nationwide agricultural extension services reaching every village.

Not all countries will be able, initially, to follow Zimbabwe's high-input approach. They may need to begin with approaches that are lower cost and which, for that very reason, offer more attractive returns. We shall be exploring many of these in subsequent chapters. They include improved seeds that can do well even with low inputs, the use of organic fertilizers and water conservation, and the wider use of trees. These purely organic approaches are capable of giving improvements of up to 100 per cent or more in the amount of useful production that can be gained from a given amount of land, at little or no cost to the farmer. They will provide the fastest, cheapest early gains, the foundation on which high-input agriculture can later be built. High-input agriculture alone can not succeed in Africa. It must have a secure and continuing base of sound conservation practices and organic inputs.

Zimbabwe's success is important as much for its symbolism as for its concrete measures. It destroys for ever the myth that Africans are poor farmers. It shows just what they are capable of, in an astonishingly short time, if they are given the right backing.

6   A green revolution for Africa

In the early 1970s, India was the focus of all the grim predictions that are now made about Africa: exploding populations, soaring food imports, impending famines. Yet within five years India began to turn the tide, and by the 1980s was virtually self-sufficient in food. That reversal of fortunes was due in large part to the Green Revolution, the spread of high-yielding varieties of rice and wheat capable of rewarding inputs of fertilizer and water with higher yields. The Green Revolution was a real revolution, a genuine break with past trends. Africa needs her own Green Revolution even more urgently than India – yet, save for a few favoured nations like Zimbabwe, she seems far removed from one.

THE LATE START

One reason is that the research effort in Africa has lagged fifteen to twenty years behind the rest of the developing world. The first dwarf wheat from Mexico was launched in the early 1960s, after almost two decades of preliminary research. The new varieties of rice were issued around 1966, six years after the founding of the station that developed them, the International Rice Research Institute in the Philippines.

Africa was a late starter. The International Institute of Tropical Agriculture was officially founded at Ibadan, Nigeria in 1967, but did not really begin work until two or three years later. IITA is the major international research centre for the humid and sub-humid tropics of Africa, and deals with crops like cassava, yams, sweet potatoes, maize, soybeans and cowpeas. Much of IITA's budget had to go on overheads like water-purifying equipment, power generators and staff housing, or was eaten up by Nigeria's ruinous rate of inflation and her grossly overvalued currency. As late as 1980, one third of IITA's research posts were unfilled.

The semi-arid tropics of Africa, which need appropriate research more acutely than any other region in the world, fared even worse. The major international research centre for the area, the International Crops Research Institute for the Semi-Arid Tropics, was founded only in 1972, to research the staple crops of the Sahel – sorghum, millet, pigeonpea and groundnut. Though most countries in this zone are in Africa, the centre was based in India. The siting probably cost the Sahel another six or seven years' delay. The main breeding effort took place in India. From the mid-1970s high-yielding Asian sorghums and millets were tested in Africa, but they were unable to withstand the continent's harsher physical conditions and distinct pests and diseases. The prospects improved when, in 1978, the Organization of African Unity set up the Semi-Arid Food Grain Research and Development Project to co-ordinate national and international research on sorghum, maize, millet, cowpea and groundnuts. It was not until 1981 that ICRISAT opened its first major research institution, the Sahelian Centre, near Niamey in Niger.

Even after the delayed start, research in Africa faced additional handicaps due to the unfamiliarity and complexity of the social and physical environment. The task was far more formidable. As Ermond Hartmans, just retiring as IITA's director general, told me: 'All they had to do in Asia was come up with new varieties of the main cereal. Here in Africa we have to do five or six things at once: developing new varieties of three or four major crops, *and* methods of managing land and conserving the soil.' Researchers began with a baggage of preconceptions based on Western agricultural practice and experience with Asia's successful green revolution. Yet almost everything was different in Africa. Most of Asia had fertile soils with good structure and a fairly reliable climate. Africa's soils were impoverished and easily eroded, and her climate was wildly temperamental. Asia's farmers grew a few staple crops, mainly in sole stands; their African counterparts grew a bewildering variety of little-known crops, and mixed them all up in their fields. Many of Asia's farmers had access to draught oxen, some capital, and plentiful, cheap labour. Africa's farmers were short of animal power and human labour, and the great majority had no spare funds at all. It took ten years and more for researchers to fully understand the special constraints of Africa.

The way research was organized worked against the chances of success. Peter Matlon, ICRISAT's chief economist in Burkina Faso, points out that conditions on the research stations where new varieties were bred were far removed from the realities of the African peasant. Stations were sited on good soils, with easy access to water even in the dry season. Their new varieties were usually selected and tested with fertilizer levels that farmers could not afford, and with high standards of ploughing, thinning and weeding that overworked peasants, mostly women, could not manage. Lo and behold, when new varieties and methods came to be tested under farmers' conditions, the results were 40 to 60 per cent worse than had been achieved on station. In the Sahelian region, the overwhelming majority of 'high-yielding varieties' performed worse under low-input farm conditions than the peasants' own varieties, carefully selected over many generations to cope with local climate, soils, pests and diseases.

THE NEW AGENDA

The lessons have now been learned, and a new research agenda has emerged. There is a growing emphasis on focusing research on the farmer's needs and real circumstances. There is more concern with carrying out baseline studies, to find out what farmers currently do, and why, and to identify the areas where research could help to improve things. New technologies and new varieties are tried out on small farms. On-farm trails managed by researchers show how new developments work out in the farmers' physical environment. These are complemented by farmer-managed trials, where farmers are allowed to try out the technology themselves, to see how it performs in their social and economic circumstances – the amount of cash, labour or draught power they have; their attitudes to risk taking; their crop priorities; their patterns of intercropping; their requirements for food, fuel or fodder. All these factors interact with the farmers' physical environment to affect performance. Individual crops and technologies can not be studied or developed in isolation, least of all in Africa. They have to be studied as part of farming systems.

The new approach builds a strong element of farmer participation into the research. The farmer's felt needs and real conditions are taken into account. There is plenty of direct feedback about how new varieties or techniques work or fail to work – feedback that is essential in fine-tuning them so that they are suitable for wide adoption. This kind of research has built up a new awareness of farmers' problems – and has begun to influence the way in which improved varieties of crops are selected. The aim is to breed varieties that perform well not only with adequate fertilizer and water and careful management, but even with low inputs, on poor soils and with erratic rainfall.

In ICRISAT's work in the Sahel, new varieties are tested under the rigorous conditions they will encounter out in the field, with no fertilizer or ploughing, with periods of drought at various stages, with sand smothering the seedlings when they emerge. They are selected not just for *good* yield – but for *stable* yield across good and bad years. High *grain* yield is not always the sole consideration – farmers in the semi-arid areas are keen to get as much fodder or fuel out of the crop residue as well.

Local varieties often correspond to all these requirements, because they have been selected over generations by farmers under just these conditions. The problem is that they do not respond well to improved inputs like fertilizer. What plant breeders are looking for now are new varieties that will perform at least as well as local varieties even with poor management, but which respond to gradually increased inputs with better and better yields, so that farmers have some incentive to increase their inputs – seeds that suit Africa's stressful environment, and don't require the environment to be changed to suit them.

Increasingly, plant breeders are working to develop varieties that mature early. These are needed desperately in the Sahel, where the average length of the rainy season has been shrinking. But they are also useful in less arid areas. They give farmers more leeway in their planting dates; they are often better suited to intercropping; and they can grow to maturity in valley bottoms in the dry season, on moisture left over from the rains.

MODEST BEGINNINGS

Africa's Green Revolution is already beginning for humid-zone crops such as maize or cassava. For these, and for cowpeas, Africa is now at the stage that Asia reached twenty years ago, with the launch of high-yielding wheat and rice varieties.

In some countries the process is even further advanced. As we have seen, Zimbabwe has had its own hybrid maize varieties since 1949. Now that these are reaching the black smallholder, Zimbabwe's Green Revolution is in full swing. In the mid-1950s, work began on high-yielding maize for Kenya. In 1964, the first hybrids were released, crosses of an Ecuadorian line with the best local variety. They gave a 40 per cent yield increase. By 1977, around half of all the smallholders in Kenya were growing high-yielding maize, though the vast majority of them were hedging their bets by growing local varieties too. Just under one third of the maize area in 1977 was growing high-yielding varieties. The seed was produced by a private concern, the Kenya Seed Company, and sold through the normal market network of wholesalers and shopkeepers. There was a considerable element of spontaneity in the spread: more than half the users had first heard of the new seeds from friends (45 per cent) or dealers (10 per cent). Kenya's hybrid seeds are now exported to Tanzania, Uganda, Ethiopia and Zaire.

Maize took off far more slowly in West Africa, where the greater humidity fostered pests and diseases. In 1975 two varieties that were resistant to lowland rusts and blight were released in Nigeria. By 1983 they were grown on 200,000 hectares, about 11 per cent of the maize area, and in Cameroon they covered more than a third of the maize area. By 1985 varieties had been developed that were resistant to additional diseases: maize streak, mottle, and stripe. All these varieties were open-pollinated, better suited to present conditions in most African countries because they keep their vigour for up to four years before they need replacing. *Hybrid* maize varieties, specially crossed by seed breeders, need replacing every year and require an effective nationwide distribution system. But in the long run, hybrids will have to be introduced: varieties have already been developed at IITA which offer yields 20–50 per cent higher than the best open-pollinated varieties.

The subsistence grains of Africa's drier areas are not so well advanced. Maize is normally a humid and sub-humid zone crop, but varieties are now in the pipeline which mature in only 80 days – as fast as millet – and are capable of giving up to 4 tonnes per hectare. A small number of promising sorghum varieties have now been developed which are proving popular with farmers. Framida is a red-grained variety which performs well even with poor management; it has vigorous seedlings and resists the scourge of sorghum, the parasitic witchweed, *striga*. E-35-1 is another high-yielding white-seeded variety which has been crossed with a number of tough local varieties. One of these crosses, labelled 82-S-90, has done well on peasants' farms. With no fertilizer or ploughing, it yielded 970 kilos a hectare – 43 per cent more than the local variety. With ploughing and a moderate dose of fertilizer, it kept its advantage. If it maintains its promise, it stands a good chance of spreading spontaneously, with or without official backing. High-yielding varieties of sorghum have been released to farms in Burkina Faso, Ethiopia and Zambia. In Sudan the variety Hageen-Dura 1, released in 1983, covered 50,000 hectares by 1985. Even in the drought year of 1984, the variety gave farmers average yields of 1 tonne – double the average Sudanese sorghum yield in *normal* years.

Millet, the grain on which the driest parts of the Sahel most depend, is furthest from success. It has to withstand the most demanding conditions – soil temperatures of up to 55°C when the seedlings emerge, followed by sandstorms of up to 80 kilometres an hour, abrading the leaves or burying them under 2 centimetres of sand. Improved varieties imported from India could not cope with the Sahel's stresses. Out of 3,000 lines tested, not one did better than the best local variety under farmers' conditions. Les Fussell, millet specialist at ICRISAT's Sahelian Centre, believes it may be as late as 1989 or 1990 before suitable millet varieties are ready for release.

After sorghum and millet, the humble cowpea is the most important crop of the Sahel. It usually sprawls along the ground between tall cereals. Being a legume, it can fix between 70 and 240 kilos of nitrogen for each hectare. The stalks and leaves are usually dried on mud roofs or in the forks of trees for livestock fodder. Cowpeas are a drought-resistant crop that will grow with as little as 200 millimetres of rain a year, and can tolerate dry periods in the growing season. Their low-spreading leaves protect the soil against erosion and help to conserve water. The main problem is that cowpea is appallingly susceptible to diseases and pests: leafhoppers, aphids, foliage beetles, pod borers, flower beetles, pod-sucking bugs, storage weevils, all take their toll. Thrips, tiny black insects that eat flowers and buds, can completely destroy the crop. These adversaries keep cowpea yields extremely low – a mere 0.3 to 0.6 tonnes of grain a hectare.

The International Institute of Tropical Agriculture began work on cowpeas in 1970. It identified varieties that resisted the various pests and diseases, but most of these were low-yielding. The plant breeders' task, then, was to cross-fertilize the resistant varieties with other varieties that produce high yield. IITA now have a number of varieties that resist up to twelve different pests and diseases, and yield around two tonnes of grain per hectare. Many of these are not the usual crawling type with pods below the canopy, but bushy, erect types that carry their beans above their leaves on strong peduncles, making insecticide spraying and harvesting easier.

IITA cowpea varieties have been adapted to local conditions and released to farmers in Botswana, Burkina Faso, Nigeria, Cameroon, Togo, Liberia, Tanzania, Central African Republic, Somalia and Zambia. One variety, TVx 3236, has spread rapidly among farmers in northern Nigeria. TVx 3236 is a cross of a thrips-resistant variety with a good-yielding local variety called Ife Brown. It was released in 1983 by the Kano State Agricultural and Rural Development Authority (KNARDA) in a massive World Bank area project, and by the following year 9,000 farmers were growing it. Farmers in contact with KNARDA's extension workers got grain yields more than three times higher than non-contact farmers. Kano's Hausa farmers renamed the variety 'Dan KNARDA' – son of KNARDA – and it proved so popular that all supplies sold out. Some farmers sold seed to their neighbours, others sent 10-tonne lorries to IITA's Ibadan headquarters, 1000 kilometres away, in a vain attempt to get more. The variety spread spontaneously, as fast as seed became available. Another IITA variety, ITA-60, matures in only 60 days – faster than millet – and is spreading in low-rainfall areas of

Kano State and around Bida, where it can grow even in the dry season, on moisture left over from the rains.

SELF-PROPAGATING SUCCESS

Perhaps the most encouraging harbinger of a potential green revolution in Africa is the spontaneous spread of improved cassava varieties from the International Institute of Tropical Agriculture, in Ibadan, Nigeria. Cassava is a root crop from South America, introduced to Africa by the Portuguese in the sixteenth century. It is now the major staple in most of Central Africa, and spreading in East and West Africa. Its bulbous, starchy roots normally mature in 12 months – but they can be left in the ground for up to two years, as a standby in case of drought. Cassava can produce two and a half times as many calories per hectare as maize, and will grow without any fertilizer in impoverished soils where nothing else will flourish. But like most other crops, cassava falls far short of its potential in Africa. The average yield is only 6.8 tonnes per hectare – 30–40 per cent lower than in Latin America. The major obstacles are pests and diseases: cassava mosaic virus, bacterial blight, and two new pests that invaded Africa from South America in the early 1970s, green spider mite and cassava mealybug.

The philosophy of IITA's root and tuber programme is a model of appropriate research. Varieties are selected under the no-input conditions that prevail on most small African farms, without fertilizers, pesticides or fungicides. They are then tested on farms, often in areas that are particularly exposed to diseases or other handicaps. Korean Programme director Sang Ki Hahn explains: 'If a variety does not do well under minimal conditions, it is not worth further propagation.'

The approach has paid handsome dividends. By 1976 one very promising variety had emerged: TMS 30572. It was much bushier than the spindly local varieties, and resistant to the two main diseases. It was ready to harvest four months earlier, and gave yields 50–100 per cent better than local varieties. It shades the soil more, and therefore protects it better against erosion and cuts down the labour involved in weeding by one third or more.

The biggest problem was how to spread the new variety, since

Nigeria had no effective extension service. 'I was told it would take a century to get the material out to the farmers,' remembers Hahn. 'In the event, it was hard to stop it: IITA junior staff stole bunches of cuttings and planted them on their own farms.' Hahn pioneered some novel distribution methods: he gave out cuttings from the back of a lorry, reaching up to 1000 people in a day, and if he saw a field of weak disease-ridden plants, he would stop and stick an improved stem in. Yoruba farmers were so grateful to Sang Ki Hahn that in 1983 they made him an honorary chief, with the title Seriki Agbe, King of Farmers.

By 1985, IITA varieties covered one fifth of the cassava area in Oyo State, and accounted for one third of the production. Nigerian farmers did most of the dissemination by passing planting material on to each other. The spread was slow, because cassava is propagated by pieces of stem, not by seeds, but it was estimated that the area under IITA varieties was tripling every year.

Oyo farmer Adejare Ajayi says he has never set eyes on a agricultural extension worker in his life, but like most African famers he is an avid experimenter and tester of new varieties. His maize, for example, is a locally developed variety with ears five times the size of older varieties. He got hold of his first IITA cassava stems from a government farm where he works as a casual labourer. He is now growing three cassava varieties on his land; he calls them *ITA Funfun* (IITA white – 30572), *ITA Pupa* (IITA red – 30555) and Odongbo, a local variety. 'When Odongbo gives two tubers,' he explains enthusiastically, 'ITA Funfun has twelve tubers clustering all around like a bunch of bananas, so thick you have to separate them with your hatchet one by one.' He wanted to plant nothing but IITA varieties, but couldn't get enough cuttings. The following year he will have enough to cover his own land, and after that he will sell cuttings to others. Of his twelve neighbours, only three are not growing IITA cassava, and then only because they could not get hold of the planting material.

SOLVING THE FERTILITY PROBLEM

No green revolution can get under way in Africa without an assault on the low fertility of her soils. But relying on high doses of chemical fertilizers is fraught with problems in Africa's present circumstances. Except in Zimbabwe, chemical fertilizers are imported, and are vulnerable to foreign exchange crises. They have to be transported and distributed, yet roads and marketing networks are poorly developed. They cost money, and the majority of African farmers have little spare cash to invest and few economic sources of credit. In many upland areas, long-term use of conventional nitrogenous fertilizers can actually reduce soil fertility by acidifying the soil and inducing potassium deficiencies. In drier areas, chemical fertilizer use can be economically irrational. In one survey in Burkina Faso, a moderate application of fertilizer resulted in a net *loss* in areas of low rainfall. Even in the wettest areas, there was a substantial risk of loss because of the patchiness of rainfall. Some 44 to 70 per cent of farmers here spent more on fertilizers than they gained in extra yield. The average financial returns – 40 to 80 per cent on the investment – were not high enough to justify this risk. Not surprisingly, farmers are sceptical of recommendations to use fertilizer. A single year's serious loss can deter them for life.

Yet there is no way of dodging the need to fertilize; as fallow periods decline, farmers cannot go on extracting crops from the same piece of land without putting something back in. In the short to medium term Africa will have to rely to the full on its own indigenous resources.

One strategy is to recycle the organic crop residues that are at present wasted or diverted to other uses. Residues such as stalks and leaves contain 40 to 90 per cent of the nutrients that crops remove from the soil. They constitute a potentially massive reserve of nutrients. In one estimate, residues in northern Nigeria contained the equivalent of 46 kilos of nitrogen, phosphates and potash for every hectare of farmland – almost 80 times the amount of *mineral* fertilizer used in the region. If crop residues are spread on the soil surface as a mulch, they have a powerful effect on crop yields and this effect is stronger in the drier areas. In the humid zone, at Ibadan for example, mulching increased maize yields by 25 to 60 per cent. In the semi-arid zone in Niger, in a drought year, millet grown with no fertilizer or mulch gave only 0.2 tonnes of grain per hectare. When 4 tonnes of mulch were added – an amount normally available in the area – the yield was 0.8 tonnes, four times as high. This was almost as good as the 1 tonne per hectare obtained by adding 90 kilos of chemical fertilizer. Mulch even improves the efficiency of chemicals: the yields of fertilized plots were doubled when mulch was added.

Mulching reduces soil temperatures. It increases the amount of rainwater that filters into the soil, and improves its ability to retain moisture – very important during dry spells. It seems to increase nitrogen fixation by free-living soil bacteria. It does require more labour than burning, but it partly compensates by cutting down the labour needed for weeding. And it protects the soil against battering rains and driving winds and reduces soil erosion.

The humid areas of Africa have massive quantities of mulch available. In the drier areas, mulching will be harder to promote because of the heavy demand on crop residues for fuel, fodder and construction. Only if trees are planted for these purposes will crop residues be freed for use as mulch.

FERTILIZER FOR FREE

The use of crop residues simply restores to the soil some of what has been taken from it. Harnessing of nitrogen-fixing bacteria actually adds free fertilizer from the air. Leguminous plants and trees have nitrogen-fixing bacteria living in nodules on their roots, and can fix anything from 45 kilos up to 860 kilos of nitrogen per hectare. That nitrogen can give a valuable boost to crops grown together with the legume, or in the same field following it. For example, when maize is grown following cowpea, the yield is more than *double* that when maize follows maize – with no added fertilizer.

The most promising use of legumes is in intercropping – growing two or more different crops intermingled in the same field. Intercropping is coming into fashion among organic gardeners in the West, but it has been second nature to African farmers for centuries. It has been estimated that 80 per cent of West Africa's farmland is intercropped. Farmers intercrop anything with everything, continually trying and testing new combinations. I have seen cassava and maize, millet and cowpea, coffee and tomatoes, cowpea and yams, cabbage and beans. The conventional wisdom used to be that intercropping was primitive and untidy. Most development projects aimed at replacing the practice with neat monocropping, which made it easier to apply fertilizers and pesticides, and easier to mechanize the harvest. But research over the past decade has confirmed that the peasants had it right all along. Intercropping is superior to monocropping in the African environment. Like mulching, intercropping provides a better protective cover of vegetation, and so lowers soil temperature, increases water infiltration, helps to prevent soil erosion, and saves labour on weeding. It reduces the incidence of pests and diseases, which cannot spread so easily when their host plants are separated by other species.

Intercropping works best when it combines crops of different heights, root depths and maturity periods, because the plants compete far less, and can take up more of the available light, water and nutrients. The total output of any area is usually greater under intercropping than under monocropping.

In semi-arid areas, when sorghum and groundnut are grown together, the combined yield is 25 per cent greater than if they are grown separately on the same area. When millet is intercropped with cowpea, the total output is 51 per cent higher than when they are grown as monocrops. Similar advantages are found in the humid areas. When cassava is grown with maize, the maize yield is often just as high as when it is grown alone, so that the cassava yields of 9 to 19 tonnes per hectare come as a pure bonus. Cassava and maize go well together: the maize matures in four or five months, giving the cassava five or six months' unshaded growth before harvesting.

Because different crops have different growing periods and maturity times, intercropping helps to even out the labour load, and spreads the availability of food more evenly, shortening the crippling 'hunger gap' each year before the main harvest. Perhaps even more important, in Africa's unpredictable environment, total yields are more *stable* from year to year and the farmer's exposure to risk is lower. Part of this stability comes from the fact that different crops react differently to climate variations. The idea was expressed perfectly by a Hausa peasant I met in Kano state in northern Nigeria. Yakubu Abdullahi was merrily growing sorghum, millet and cowpea in the same field. 'As a camel's back can carry many burdens,' he explained, 'so the earth can carry as many crops as we plant in her. If there is rain here, we get sorghum. If there is no rain, we get millet.' But part of the stability seems to come from a sort of compensatory effect: if one crop is weakened, whether by drought or by pests, the other has more elbow room to grow. ICRISAT has calculated that the probability of disastrous crop failure when pigeonpea is grown alone is one year in five; when sorghum is grown alone, one in eight; when both are grown, but in separate plots, the chance of failure falls to one in 13. But when they are intercropped, the failure rate drops to only one year out of 36. Intercropping is a way of making the most out of limited resources. The more limited the resources, whether of water or of nutrients, the greater its advantage. When crops at ICRISAT got 280mm of water, intercropping had a 20 per cent advantage in total output over sole cropping. When crops got only 40mm, the advantage soared to 60–160 per cent.

African farmers are among the world's greatest experts on intercropping. But science can help improve their practice by testing out different combinations of times of planting, maturity periods, dates of harvesting, densities and arrangements of plants. It must also take intercropping into full account in its own development of new breeds and techniques. African farmers will go on intercropping for decades, and new varieties – if they are to be widely adopted – must be adapted to intercropping. For example, in western Nigeria, where maize and cassava are grown together, a high-yielding maize that results in seriously lowered cassava yields will not prove popular among small farmers. An early maturing maize with erect upper leaves that does not shade the young cassava bush too much will stand a better chance of adoption. Varieties can be specifically bred for intercropping by emphasizing complementary plant build. A dwarf sorghum goes best with a tall millet. A low-spreading cowpea grown with a tall type of maize yields 66 per cent more than a bushy cowpea with a short, stocky maize.

Research can also help to increase the amounts of nitrogen fixed and the share available to boost associated crops, breeding improved strains of legumes and associated rhizobia, developing patterns of intercropping and rotations that make the maximum use of nitrogen fixation.

Improved use of crop residues and nitrogen fixation can take African farmers a long way. They could boost their very low yields without any outside inputs. In the long run, if Africa's massive increases in population are to be accommodated, the use of chemical fertilizers will have to spread. But the type and quantity will have to be much more carefully chosen than in the past, better adapted to soil and rainfall patterns. Calcium ammonium nitrate, for example, is less likely to acidify the soil than urea. Perhaps the most promising additions are phosphate fertilizers. Quite modest applications can produce spectacular yield increases. The reason for this effect may be that phosphates, which are deficient in most African soils, stimulate root development. Deeper roots allow plants to exploit deeper moisture levels, and therefore act as a protection against drought, or dry spells within the growing period. In Niger, for example, an application of only 15 kilos of triple superphosphate doubled millet yields from 0.55 to 1.1 tonnes per hectare. A mere 10 kilos per hectare *tripled* the yield of cowpeas. The effects of one application persist. The year after the application, cowpea yields were still double those of untreated plots. The Niger farmers who took part in the trials applied phosphates to 5 per cent of their land. They were so impressed with the results that within three years they were applying phosphates to 60 per cent of their fields, all on their own initiative, with an enterprising Hausa merchant bringing in the supplies.

The achievements are mounting, but the research network in Africa still needs more resources and a more rational organization. There is a strong argument, for example, for creating a specialized international centre for the semi-arid tropics just for Africa, perhaps by up-grading and expanding ICRISAT's Sahelian Centre. The existing pattern of research is still too focused on single crops. African farms are complex organisms, in which soils, climate, people, crops, trees and livestock interact. Trees, as we shall see, are crucial to the future of Africa's environment, yet they are not officially part of the brief of either of the two main centres. Livestock is crucial, yet research is hived off into a separate institution, the International Livestock

Centre for Africa, in Addis Ababa, while research into animal health is focused on yet another organization, the International Laboratory for Research into Animal Diseases, in Nairobi, Kenya. The ideal arrangement would be to have three international centres, one for the humid tropics (IITA), one for the semi-arid tropics (an African ICRISAT), and a new one for the highland areas. Each of these centres would integrate research into *all* crops, trees and livestock within their zones, aiming to develop integrated systems that could build on and improve what farmers already do.

International research, of course, can only have significance if it feeds into African national programmes. Africa's sheer diversity, both on the large scale and the small, means that all new varieties and new techniques have to be adapted to national and local circumstances. National research programmes in Africa are weak, despite the fact that Africa spends a higher proportion of her agricultural income on research than Asia. But much of the money is misdirected or ill-spent. The emphasis on cash crops, dating from colonial days, still persists. New programmes for food crops have been grafted on gradually, as separate activities, rather than shifting the whole focus of programmes. Even as late as 1982–3, the total number of professional research staff for sorghum and millet in sub-Saharan Africa was about equal to the number of wheat and barley researchers in Egypt. Togo, Congo, Gabon, Sao Tomé, Angola, Central African Republic and Equatorial Guinea had not a single professional researcher working on cereals. Aid agencies have funded the building of new facilities, rather than the improvement of old ones, and the normal aid project lifespan of three to five years is far too short for significant results to be reached.

Even national research only has significance if it feeds into nationwide extension, supply and credit services, capable of getting out to small farmers the results of research, the new seeds, inputs and equipment needed to get the best out of them, and advice on how to manage them and adapt them to local circumstances. The relationship between research and extension must be a two-way street. The extension service can act as a vast network of on-farm research, feeding farmers' concrete experience with innovations back to scientists, pointing out problems and obstacles to adoption.

With such national networks in place, Africa's green revolution could begin to make a real impact on a wide scale during the 1990s. In the first half of the decade new varieties of cassava, upland rice and maize will have spread to a large enough acreage to make in-roads into the massive food deficits of the more humid parts of Africa. It will not be until 1995 or later that the revolution spreads widely in the Sahel, first with new varieties of cowpea and sorghum and then of millet. In countries that do not have effective extension, credit and supply systems, and pricing policies that give farmers an incentive to produce more food, the process will take far longer.

Africa's green revolution can not be the same as Asia's. *It must, at least to begin with, be low-cost, relying on the very cheapest technologies that provide the highest pay-offs at the least risk*. It must, again to begin with, depend as little as possible on imports. Later, as farm incomes and national incomes build up, higher-cost approaches, involving imports, can gradually be adopted. What these progressive stages might involve will be spelt out in more detail in Chapter 16.

But because of Africa's sensitive environment, her green revolution must, at all stages, be built on a solid foundation of conservation, of soil, water and trees. It is to these that we now turn.

7 Gaining ground: success in soil conservation

Land degradation takes insidious forms in Africa, moulding landscapes of the kind you see in temperate regions only in the loose spoil-heaps of mines, or clay and shale coastlines worn away by the battering sea.

In the hilly semi-arid Machakos district in Kenya, sheet erosion has stripped the topsoil from many slopes, leaving hard gravel dotted with sparse, flattened starlets of grass. At imperceptible dips, where the rainwater concentrates, shallow rills form. Where larger amounts of water flow together, on the edge of paths or poorly drained roadways, the rain carves gullies, like jagged wounds torn by a rough blade in the earth's flesh. In Nigeria, patches of soil have turned literally to stone as underlying layers, rich in iron and aluminium, have hardened rock-solid on exposure to the air.

There are no comprehensive surveys of the scale of erosion in Africa, but scattered reports build up an alarming image. Croplands in Madagascar are losing from 25 to 250 tonnes of topsoil a year from every hectare. In Zimbabwe half the communal lands are severely eroded, with annual topsoil losses of up to 50 tonnes a year. Even on gentle slopes of 1–2 per cent, soil losses of 30–55 tonnes per hectare have been reported in West Africa.

Topsoil forms by the slow interaction of plants and rocks over millennia. Each centimetre takes more than a hundred years to form. Yet at the erosion rates commonly found in Africa, the work of a century is being washed or blown away in a mere six or seven years. In Lesotho the average soil depth has declined from 38 centimetres to 28 centimetres during this century, according to one estimate.

Africa is perilously sensitive and susceptible to land degradation. A 1979 study by FAO and UNEP found that 36 per cent of the land area of Africa north of the Equator is susceptible to some degree of wind erosion, and 8 per cent to dangers of salinity. 17 per cent is liable to water erosion under normal vegetation cover – but virtually the whole area is liable when its natural cover is removed. The only major area free of risk is the heart of the Sahara desert. Virtually no inhabited area is unaffected, and many areas face potential soil losses of more than 50 tonnes per hectare. Such risk levels from water erosion are found over large tracts of Sierra Leone, Liberia, Guinea, southern Ghana, Nigeria, Zaire, Central Africa Republic, and Ethiopia. Wind erosion poses similar levels of threat to much of Senegal, Mauritania, Niger, Chad, Sudan, Somalia and northern Ethiopia.

The potential consequences for food production are grave. Erosion removes the top layers of the soil first. These are richest in organic matter and in nutrients of all kinds – so erosion reduces soil fertility and cuts plant yields by an estimated 2–3 per cent for every 10 tonnes of soil lost per hectare. Soil depth is reduced: plants cannot root so deeply and are more vulnerable to dry spells. Where the soil is silty, crusts form when vegetation is removed and keep all water out of the soil. Run-off increases, carrying a heavy load of soil particles. Downstream, dams and irrigation canals silt up and their efficiency is reduced. Floods increase.

No less than 130 million hectares of Africa's total cultivable area of 789 million hectares could be lost to food production if no conservation measures were carried out, according to another FAO study published in 1984. The total productivity of rain-fed cropland would be cut by as much as 29 per cent. In fifteen African countries, uncontrolled soil degradation would cut by more than half the number of people that could be fed from local production (see map, p. 352). Most of these countries are in the drier areas of the Sahel and Southern Africa and the hilly areas of Eastern Africa.

As we saw in Chapter 2 (p. 36), much of Africa's rain falls in powerfully erosive cloudbursts. Many of her soils are poor in clay and organic matter, and are easily eroded. Yet under natural circumstances the environment is stable. The guarantee of that stability is vegetation, acting like a coat of chainmail between the attacking elements and the earth's vulnerable skin. Grasses, shrubs, trees, all break the force of wind and the impact of rain, slow down their flow, and allow water to infiltrate gradually. Vegetation not only prevents erosion, but keeps more moisture in the soil for plants to use.

The key to environmental stability in Africa is maintaining the protective cover of vegetation. Traditional agriculture did just that as long as population densities were low. Most of the land was under fallow bush and tree cover. The areas that were actually farmed were tilled only shallowly — traditional hoes and ploughs did not permit otherwise. Roots and stumps were left in place and anchored the soil — again, less out of conscious choice than because removing them would have been too laborious. Intercropping was virtually universal, and provided a longer and denser cover for the soil than sole cropping. These methods were not adopted with the prime aim of conserving the soil: there was little conscious thought of the need for that. But conservation was the end result.

Those tribes that farmed in hilly or more densely populated areas did have to think actively about conservation, and many developed their own effective and appropriate methods, from the terraces of the Dogon or Konso, to the soil pits of the Matengo or the multi-storey farming of the Chagga (pp. 74–78).

The causes of soil erosion in Africa are therefore all those forces that exacerbate the removal of vegetation cover, plus those that stand in the way of alternative methods of conservation. Rapid population growth, and the parallel growth in animal numbers, has led to overgrazing, reduction in fallow areas and felling of trees. In most places land is still abundant enough for it to be easier and cheaper to move on to a new patch than to invest time and effort in rescuing an old degraded patch. Official neglect of agriculture and low prices paid to farmers have provided no resources or incentive to conserve the soil. Misguided official development programmes have pushed monocropping instead of intercropping, deep ploughing instead of shallow, removal of roots and stumps by bulldozer. Many of the traditional soil conservation methods began to decay: they relied on heavy inputs of labour, and as the demands of colonial mines and plantations grew, and later the expansion of cities with their better wages, the shortage of labour in rural areas intensified. Finally, in the Sahel, all those long years of low rainfall played their part in weakening vegetation cover.

Soil erosion first became the subject of serious concern in British East Africa as early as the 1930s. But instead of educating and encouraging African farmers to conserve their land, the British authorities tried to impose conservation by force. Owners of sloping land were compelled to build elaborate terraces, and to reforest the steepest slopes. These measures were expensive, or involved a sacrifice of cropland to woodland at a time when there was virtually no market for wood. Offenders were fined, whipped or jailed. The conservation programme was one of the most intrusive and unpopular aspects of British rule, and helped to fuel the Land and Freedom ('Mau Mau') rebellion of the early 1950s in Kenya.

Since independence only a handful of African governments have given the fight against land degradation the priority it demands. When large-scale efforts were mounted, they were mainly high-cost, import-intensive projects that did not involve the local population. The emphasis was on physical structures like terraces and bunds (low earth ridges), painstakingly designed by experts or expatriates and built by tractors and bulldozers. Most of these works were not maintained. Many were deliberately destroyed by the supposed beneficiaries.

The massive programme mounted in the early 1960s by the European Development Fund in Upper Volta (now Burkina Faso) was typical. Bulldozers covered 120,000 hectares with a network of low soil bunds. There was no involvement of villagers whatsoever: peasants remember only the massive earthmovers trundling uninvited across their land as if alien invaders had landed. There was no explanation of the purpose of the bunds, no training of villagers in how to maintain them. The bunds held back water uphill and allowed it to infiltrate slowly, but farmers complained that they dried up the soil for a few metres below the bund. Follow-up studies found that most bunds were not maintained at all, and decayed in two or three years. Farmers left their old footpaths or cattle tracks in place crossing the bunds, creating holes through which water dammed up behind would surge, wearing a gully below. Many of the old bunds, worn to low bumps in the ground, can still be seen meandering across cropland that has now degraded to virtual desert: a potent testimony to the failure of the whole venture.

SAVING THE LAND BY SELF-HELP

Successes in soil conservation are few and far between. Perhaps the most outstanding programme in Africa has been Kenya's.

The bulk of Kenya's population lives in hilly areas, highly vulnerable to rain erosion. The British authorities saw the danger, but their high-handed, coercive approach did more damage than good. Conservation became synonymous with colonial oppression, or collaboration. After independence, many farmers let their terraces and drains decay, or even destroyed them.

More than a decade elapsed before Kenyans overcame the negative attitudes instilled by the British. The shift began early in the 1970s when Kenya's national report to the UN Conference on the Environment in Stockholm underlined the gravity of the soil-erosion problem. Kenya approached the Swedish government for help, and in 1974, with finance from the Swedish International Development Authority (SIDA), the soil conservation programme was launched. Within a decade, no less than 365,000 Kenyan farms – two out of every five – had been terraced under the programme. Annual targets were regularly and handsomely over-fulfilled: 62,000 farms were terraced in 1983–4 – 42 per cent more than the target of 44,000. Many more farmers, especially in the densely populated highlands, were carrying out thorough conservation measures under their own steam.

The programme's success owed much to the approach of Sweden's Professor Carl Gösta Wenner, who was seconded to Kenya in 1974. Wenner began by asking farmers about their problems, needs, and existing methods of soil conservation. He found a number of effective traditional methods in use, and worked with farmers in a series of regional projects to improve and perfect them. The end result was a set of measures that

Fanja juu or 'work-up' terraces level themselves over a number of years. The ditch can be used for bananas or pawpaw, the leading edge for perennial fodder grass

farmers could and would carry out themselves, with their own resources and a modicum of expert help.

Three main methods of terracing were developed. *Fanya Juu*, or 'work-up' terraces, were based on the techniques of the Kamba people of Machakos district. This method involves digging out a narrow trench, and throwing the soil uphill to form a ridge. Trash lines were used intensively in the Taita hills. Here weeds, stalks, twigs and other refuse are piled into long lines along the contour of the hill. In the traditional system, this was a method of composting and the lines were moved each year. Under the SIDA programme, the lines are left in place, perhaps anchored by branches, and gradually collect soil and other debris. In both systems, the ridges are stabilized by planting perennial fodder grasses, or trees for fodder, fuel or fruit. In the course of a few years, soil from the back of each section of slope is washed towards the front, and a series of level terraces forms automatically, with no heavy work required other than routine maintenance of terrace edges. The third method uses long infiltration strips along the contours, also planted with perennial grasses, where rainwater can filter into the soil, greatly reducing the amount of run-off that gathers and accelerates on cultivated slopes.

The measures are simple and easily mastered. Only hand tools are needed – no expensive imported machinery, vulnerable to lack of spares or maintenance. The programme, therefore, is proof against foreign exchange crises. The cost is low. Labour and the renewal of worn-out hand tools are the main expense for farmers. The work can be carried out in the dry season, when it does not conflict with food production. No cropland is lost – indeed, land is gained, for without conservation works much land would have to be fallowed frequently. The cost to government is also modest – around $30 per farm terraced in 1983–4.

The pay-offs are rapid and handsome. The terraces not only halt the loss of topsoil and the gradual decline of yields; they can actually *increase* yields, from the first year in the most vulnerable soils. This happens because more rainwater filters into the soil for crops to use, instead of running off, and fewer nutrients are washed away. With increased crop growth, there is more organic matter, more root growth and more earthworm activity, so the structure and fertility of the soil improves. These benefits are all the more pronounced the more arid the land is: in drought years, terraced land is more likely to produce a crop than unterraced land. SIDA studies have found that maize yields of farmers with conservation measures are 60–100 per cent higher than of those with none, and bean yields 80–130 per cent higher. Assuming that half the advantage is due to improved farming, a conservative estimate suggests that terracing boosts crop yields by 30–65 per cent. When terrace edges are planted they produce an additional bonus of fodder, poles and wood. Finally, terraced land is flatter and easier to cultivate than slopes. Overall, the return for a day's work on terraced land is 39–47 per cent higher than on unterraced land. Added to this is the incalculable value of preserving the soil for future generations.

Farmers participated in the design of the technology, ensuring that it was adapted to their needs and resources. They also participate in the making and maintaining of conservation works. Government extension workers plan the layout of the terraces, but farmers themselves do the hard work and bear the bulk of the expenses. Until recently grasses and trees for planting along terrace edges were free; now farmers have to pay a subsidized price. The government does pay for certain communal works, mainly cut-off drains to channel rainwater flowing off roads and watersheds, away from farms downhill. For the most part, these paid works are done by local farmers – often single mothers working for a daily wage. In Machakos, where the Kamba people have a long tradition of collective self-help, there is no payment for communal work.

Kenya has a very effective system for spreading soil conservation. The basic work at local level is done by technical assistants – all-purpose agricultural extension workers. They advise on the measures that are needed, and plan the layouts. They can call on a back-up team of specialized soil conservation officers at district level. For most of the dry season, the technical assistants concentrate on soil conservation work, shifting to crop production in the rainy season.

This national network of advice and support meshes at local level with a mass of popular organizations capable of mobilizing people. The bulk of soil conservation work is done by women's groups. 'Women are the backbone of soil conservation,' said Joshua Owiro, the conservation officer who showed me round Machakos district. 'Without women, there wouldn't be any soil conservation in Kenya.'

Machakos is a semi-arid area south-east of Nairobi. It was the site of the most rapid and serious degradation in the country, and the effects are still visible: deep rills in fields where oxen stumble when ploughing; deeper gullies like fissures opened up by earthquakes, cutting off one house from another; grazing areas stripped down to rock or gravel. But the most impressive sight for the traveller to Machakos today is the terraces. There is hardly a single farm without them, and the vast majority are well made and well maintained.

Overlooking the town of Machakos is a steep hill called Kimakimu. It is cleared of trees and farmed right up to its summit. Rainwater floods down from the upper slopes, coursing into two gullies that have become chasms into which children and livestock can fall. So much topsoil had been washed away, and so much water and nutrients were being lost every year, that maize yields had fallen as low as a quarter of a tonne per hectare. In 1981 the Kaluoki women's group began an ambitious series of conservation works on the hillside. By 1985 all but a handful of farms were terraced, and the whole hillside was notched with zigzagging cut-off drains to channel rainwater away from the fields. Local women say maize yields have doubled since the work was done.

Across some of the deeper gullies, the women have built dams of piled rocks held in place by cages of wire netting. Behind these check-dams, earth has already begun to pile. Deep gullies of this kind can never be completely recovered, but they can be stopped from getting worse, and used as water channels for major flows. The best cure for gullies is prevention. Small gullies at Kimakimu have been planted with sisal, which will hold the sides and prevent further erosion. The network of terraces and drains should prevent further gullies from forming.

Kithuya women's group is based at Kiteta village, under the towering Mbooni Hills in Machakos. The group was formed in 1979. Like most of Kenya's 6,000-plus women's groups, it provides mutual support among women who have to shoulder a work burden that would break many backs. The forty members set up their own tree nursery and built a new primary school. They formed a savings club that enables members to buy tin roofs or bricks, and helps them out if they have an accident or a severe illness. One day a week, in the dry season, members come together and work for six hours on communal conservation work. The day of my visit they were working to rescue a valley scarred with gullies, digging cut-off drains and erecting fences to keep out grazing livestock. The women's group members spend another day each week terracing the plots of individual members. Teresia Wambua showed me the ten deep *fanya juu* terraces, each one 60 metres long, that the group had just built on her one-hectare plot. Teresia would never have managed it alone. Her husband works in Nairobi, and she spends three hours a day fetching water. A single woman would have taken 150 days to do it. The women's group completed the terraces in just two days.

The productivity of the group is much greater than the sum of its members. They raise one another's morale and keep up their pace with rhythmic songs. They hold one another to fixed dates and targets that a woman alone, faced with all the demands on her time, could so easily be tempted to shelve.

Kenya's soil conservation programme lays heavy emphasis on training. Each year around 1700 technical assistants are trained or retrained. Training is also aimed at administrators: 200 district officers a year, and 1,000 chiefs and sub-chiefs, have two- or three-day seminars and field trips on soil conservation. This ensures that the whole government structure, not just specialist staff, understands and supports the soil conservation programme. Soil conservation will soon be integrated into the school syllabus. All Kenyan secondary schools have their own farm plots, and 1,000 teachers a year are trained in soil conservation.

An important factor in the success has been the Kenyan government's powerful commitment to soil conservation. The message is put across through every available channel. Ceremonial occasions play their part in an orchestrated campaign. Once a year, shortly before the long rains in April, there is a National Soil and Water Conservation week in which every locality undertakes some major project. Each year one of the country's provinces selects a very badly degraded site. Here Kenya's President, Daniel Arap Moi, and senior politicians put in not just a token appearance but a whole day's hard labour with lavish media coverage.

A more continuous expression of commitment is the Presidential Commission on Soil Conservation and Afforestation. Set up in 1981, the commission co-ordinates the many government agencies, donors and voluntary groups involved. It reviews performance, policies and legislation, and recommends new measures and projects. It is the only such commission in Kenya, and its existence places conservation in a central role, embracing and transcending all the specialized ministries. Recently the commission's work has been strengthened by the creation of Soil and Water Conservation and Afforestation Committees at every level of government from district to locality, with a similar role in co-ordinating and selecting priority activities.

Kenya's soil conservation programme is a model of responsible aid. A single major donor is involved in the nationwide programme, so there is no confusion of approaches, no diversion of African energies in dealing with multiple aid missions. SIDA's commitment is not for the usual four- or five-year span of aid projects. It has already lasted for eleven years, and will continue for at least another five. Assurance of funds over such a long period protects soil conservation in Kenya against the budget crises that are so frequent in Africa. The bulk of SIDA's aid goes to pay local costs, and the import content is minimal. Where many projects set up separate staffs to make sure they get results, SIDA's programme is fully integrated into the Ministry of Agriculture and aims to strengthen its capacity. Where many projects pick on a single district and give it hothouse treatment, the SIDA programme is nationwide.

As always, there are potential improvements that could be made. Paying farmers to dig cut-off drains to protect their own fields is not only a major expense; farmers feel the cut-off drains are the government's responsibility, and maintain them less well. Using unpaid groups of farmers whose land was protected by the drain would cut the cost and might improve the chances of long-term maintenance by local people. There is a tendency to go for more elaborate terraces where cheaper methods like trash lines would do just as well. Methods should be more frequently modified to suit local needs, reduce labour costs, and build on local traditional methods. Perhaps the most significant drawback is the failure to use trees on a wide enough scale. Closely planted hedges of leguminous fodder trees along terraces would produce more fodder and do a better job of soil retention than grasses. The integration of agroforestry into soil conservation should be the next priority in Kenya.

ETHIOPIA: STAUNCHING THE HAEMORRHAGE

On the wall of the Ethiopian Ministry of Agriculture's conservation office in Harar, two large photographs hang side by side.

The first shows an octopus network of gullies extending its tentacles across eroded soils; the second a stepladder of terraces carrying healthy green strands of crops. 'What is the difference between these two photographs?' the caption asks. The answer, 'Human labour'. The amount of human labour mobilized for conservation in Ethiopia is probably unrivalled in Africa – currently around 35 million person-days each year, or the equivalent of a permanent workforce of 135,000 people working five days a week.

No country in Africa needed an effective conservation programme more urgently. Seven out of ten Ethiopians live in the mountainous highlands, where three fifths of all land has slopes over 16 per cent. According to recent studies, more than half the highland area shows signs of accelerated erosion. Each year it is estimated that 3.5 billion tonnes of topsoil are washed away, an average of 70 tonnes per hectare, and 100 tonnes a hectare from cultivated land. Indeed annual losses of almost 300 tonnes a hectare have been recorded in the case of sloping fields ploughed to a fine tilth for the tiny-seeded traditional cereal, teff. The typical Ethiopian farm household of six now harvests less than a tonne of grain a year – less than enough for the barest subsistence diet – and crop yields are steadily declining due to erosion and exhausted fertility.

The causes begin at the top of the hill. Deforestation of slopes has proceeded at a ruinous pace. In 1900, Ethiopia's forests stretched over 48 million hectares. Today they are down to 4.7 million – an average annual loss of half a million hectares. As a result, rain no longer infiltrates, but sheets downhill. The pressure of population has forced farmers to cultivate steeper and steeper slopes – in one survey two out of five farmers cropped land that was all, or mostly, hillside. Some farmers traditionally build simple soil bunds on the steepest land, but these rarely withstand each year's downpours.

Deforestation has led to a critical shortage of fuelwood – the sustainable yield now meets only just over a quarter of annual consumption. The shortfall is made up partly by the felling of trees in the dry savannah of the Rift valley, by impoverished nomads with no other source of cash income, and partly by burning crop residues and dung. The soil, deprived of all organic residues, becomes even more vulnerable to erosion.

The outlook is alarming. Already about 20,000 square kilometres have soils so shallow that they can no longer sustain cropping. If present trends continue, the Ethiopian Highlands Reclamation Study forecasts, that area will quintuple over the next 25 years to 100,000 square kilometres. If degradation is not checked, average incomes in Ethiopia will be 30 per cent lower after 25 years. 'Today's children,' the study warns, 'could see over a third of the highlands become incapable of sustaining cropping, while the population trebles in their lifetimes.'

Ethiopia is the site of the World Food Programme's biggest conservation effort in Africa. Like most WFP projects it provides food for hungry people not in the form of free handouts, but in exchange for hard work aimed at laying the foundations of food self-sufficiency. The project began in a small way in the northern provinces of Eritrea, Tigre and Wollo, after the 1973 drought. In 1980 it was expanded and now covers 44 densely populated catchment areas seriously affected by drought, land degradation and food deficit. The initial programme was ambitious enough, involving 44 million person-days' work, to be carried out over a period of four years. The target was reached in only two years. In 1982 the programme was expanded again, and has now been extended up to 1989.

Harerghe was one of the first regions where large-scale conservation work began. The Chercher highlands of Harerghe reach out eastwards from the main Ethiopian massif, sandwiched between deserts to the north and south. Within living memory they were covered in dense native forests of conifers. Isolated pockets still remain, but most have been cleared for farming or fuel. In many valleys the patchwork of plots reaches right up to the crests of precipitous hills, sometimes on 50°–60° slopes. Gullies eat into the lower slopes, while many of the potentially fertile valley bottoms are too waterlogged to grow crops, and are used for grazing in the dry season.

Each catchment is an organic whole: the damage that occurs on the lower slopes is inseparable from the state of the upper slopes and watersheds. Perhaps the most impressive feature of the Ethiopian programme is the way that treatment is planned and executed for the whole catchment, in an integral fashion combining forestry, soil conservation and water development.

The small town of Hirna nestles in the fold of one such catchment. The hill that rears up behind it testifies to the former state of things. Only a thin cap of forest remains, then a shoulder of farms, now worn away to the yellow subsoil, then a row of steep outcrops. In between these, crevices carry streams in the rainy season: below each one a gully begins and deepens till it joins a monstrous super-gully, 15 metres deep, that snakes for no less than 9 kilometres the whole length of the valley.

The villages that make up Oda Bellina Peasant Association are cut off from the town and from each other by this Hydra and its heads. The energetic chairman, 37-year-old Ibrahim Abdi, explained how things came to this pass. Before the 1974 coup, which put an end to feudalism in Ethiopia, the whole of the valley belonged to ten landlords. The peasants were semi-serf labourers. If anyone was dismissed, they moved uphill to farm the higher slopes, and became share croppers, giving one third of their meagre yields to the lords. 'The first year of cropping on the hills was all right,' Abdi explained, 'but every year after that the yield would drop. After four years, we'd have to leave it fallow. If we used it again, it would be completely finished after two years.'

The 1974 revolution took the land away from the feudal lords and vested it in the Peasants' Association. This allocates land for individual farmers to use, though they cannot sell, rent or inherit it. The forests became state property – yet, paradoxically, the revolution speeded up deforestation. 'In feudal times the big trees were cut only for the landlords' houses,' says Abdi. 'After the revolution we wanted bigger houses, and we felt we could use the trees as we liked. We thought we must use them for our needs. We didn't think about the consequences.'

The conservation programme set the process of degradation into reverse gear. Land with slopes above 35 per cent was stone-terraced and planted with eucalyptus. Some 25 or 30 farmers were cultivating these slopes – light-coloured patches under the trees mark the site of their old plots. Under a system of private land-ownership, as in Kenya, it would have been extremely difficult to dislodge these hill-farmers, while protecting their livelihoods. Under the Ethiopian system it was easier. They were allocated new land lower down, by shifting around everyone else's holdings. They were also compensated with World Food Programme wheat – 0.6 tonnes for every hectare they lost, roughly the yield they would have obtained – to tide them over the first year till their new plots started producing.

The Peasants' Association leaders, convinced of the need for the shift, called a general meeting. With the help of conservation workers, they convinced the rest: 'Everyone agreed that unless those people came down from the hills, we couldn't reforest them,' Abdi remembers. The replanted slopes were closed to grazing animals, but people were allowed to cut forage from under the trees, to take to their livestock. When the eucalyptus have matured, the villagers will be able to cut them for fuel or poles on a managed basis.

Oda Bellina farmers have built soil bunds on about a third of their farmland – low terraces about 50cm high. These will be raised by 25cm each year as soil accumulates behind them, until level bench terraces form. After the hillsides were planted and terraced, villagers attacked the 9-kilometre-long gully. Check-dams of stones were built to slow the water flow, and tall elephant grass was planted in the bed along with *Acacia cyanophylla*, a weeping leguminous tree that flourishes on degraded soils. The banks were protected by lines of eucalyptus. What was once a useless ravine, corroding the cropland around it, has become a source of wood and fodder. As the final act of the comprehensive catchment plan, a small earth dam was built near the town to collect the run-off from the hills. It now irrigates some 60 hectares of rich dark soil throughout the dry season, producing handsome crops of cabbage, onions, potatoes and sweet potatoes. This more than makes up for the impoverished hillside patches now reforested, which would have had to be abandoned within a few years in any case.

The physical achievements of Ethiopia's WFP-backed conservation programme are impressive. Up to 1985, some 700,000 hectares of land had been reforested or terraced. Almost 20,000 hectares of forest are planted each year, and no less than 200,000 kilometres of soil or stone terraces constructed. In a continent where inaction is the norm, and in a country where no conservation work at all was done until the early 1970s, the programme must be accounted a success as far as it goes. The reasons for that success are partly implicit in the World Food Programme's manner of operation. It can provide only food, plus (in Ethiopia's case) part of the cost of transporting it inside the country. It does not provide machinery, tools or expert manpower for supervision. Therefore the conservation techniques chosen are of necessity labour-intensive, capable of being carried out with the farmer's available tools – hoes, shovels and traditional picks. They have to be simple and not require much in the way of technical supervision. For levelling terraces, a simple system is used. A 10-metre length of string is stretched tight between two poles marked at intervals of 5cm. A spirit level is used to check that the string is level.

Government backing was strong. Soil and water conservation have an important place in Ethiopia's ten-year plan – indeed, for drier areas they are now seen as the most reliable approach to increasing food production. And donor commitment has been consistent. The project has been running with the same conservation focus, but on an ever-expanding scale, for twelve years, with at least another three scheduled. The participation of the elected Peasant Associations was crucial. They acted as intermediaries between their members and the experts, mustering and organizing work parties and food distribution.

Studies are still under way of the impact of the measures on soil conservation and crop yield. Soil losses may be cut by as much as 60 per cent, and with that the inevitable decline of yields by 2–3 per cent a year will be averted. Indeed, yields may actually increase by 10–20 per cent where no fertilizer is being used, and by up to 50–60 per cent where it is (since less fertilizer is washed away). But this benefit depends on rainfall, soil type and slope: it is very pronounced in drier zones, but less so in rainier areas with heavy clay soils that hold a lot of water. On really steep slopes, more land is lost to terracing and the benefits take longer to build up.

Food aid is often criticized on the grounds that it deters local food production by holding down food prices, and shifts local tastes towards cereals like wheat that cannot be grown locally. The charges are valid for much food aid to Africa – especially bilateral food aid given directly to governments – but they do not stick for the WFP food-for-work aid to Ethiopia. The Ethiopian highlands are one of the few places in Africa where wheat can be grown, and it is already grown on a wide scale. A 1983 review found that there was an overall food shortage in the areas where the project operated. Food aid had only small and short-term effects on local prices, certainly not enough to affect the subsistence farmer's top priority of producing enough to meet his own family's food needs.

In a country like Ethiopia, food aid or cash aid with which peasants can buy food is unavoidable. The alternative is mass mortality from starvation. WFP payment is on a piecework basis, averaging 3 kilos of wheat plus 120 grammes of vegetable oil per day – enough for a day's subsistence diet for a family of six. The payments amounted to two thirds of participants' own production in 1983, and seven eighths in the worst drought year of 1984. Three out of five families said the food aid had saved them from starvation.

Food-for-work is far better than free relief food. It does not create a feeling of dependence and beggary. According to social surveys, participants feel they are supporting their families by their labour. The food payments are regarded as a sort of wage, which people prefer to cash payments because of the shortage of food on the market. Relief food results in the mere survival of beneficiaries; it does not help them towards self-sufficiency, nor does it protect them against a recurrence of famine. Food-for-work creates tangible assets – forests and terraces that will sustain food and fuel production in the future.

It would be churlish to cavil at the project's impressive achievements. Nevertheless, it does suffer from a number of serious flaws, some of them inherent in the very nature of food aid. By definition this consists of imports and import dependence always makes projects vulnerable. Hitches and delays can arise at any stage, from the procurement of food through its transport by sea and distribution by land. In 1984 the massive influx of famine relief seriously hampered the programme's vital work. The Ethiopian government gave priority for transport and storage to emergency relief food, as hold-ups might have jeopardized the goodwill essential for further deliveries. Food destined for conservation workers was held up, and work programmes ground to a halt, in Harerghe and Gondar. Some 50,000 workers were laid off – and almost immediately an additional 300,000 people, equivalent to the workers plus their families, began taking relief food.

The food, of course, acted as an incentive to farmers to start conservation work, but at the same time it confused the economics of the situation. Free food considerably reduced the cost of all operations. Thus there was no need to go for the most cost-effective methods. Over-elaborate measures were often applied where cheaper approaches would have been just as effective. Stone terraces were used to collect moisture for hillside afforestation, where soil basins round each tree would have done the job just as well. Whole catchments were laboriously planted with seedlings, when in many cases natural forests would regenerate within a few years as long as grazing animals were kept out. Terracing was used on fields, where grassed strips along the contour might have served better, producing fodder instead of losing land.

Conservation programmes that do not pay farmers have to work much harder at motivation and education. Farmers have to be convinced that conservation is essential before they will invest their sweat capital. Once convinced, they will do the work thoroughly and make sure it is well maintained. When they are paid to do conservation work, the results are at times perverse. The payment is incentive enough – so education does not need to be so intensive. But then the lessons are not learned so well. For example, less than one in five terraces in Ethiopia are planted with forage or trees, as they should be. Few farmers spontaneously build terraces on their fields. They tend to wait till they are included in the food-for-work schedules, since terracing ahead of time might reduce the amount of potential work-for-food in the area. Some peasants have developed work-to-rule mentalities and are unwilling to do conservation work

unless paid in food for it. Maintenance is often poor, terraces
that have been breached are not always repaired, and terrace
levels are not always raised each year as they should be.

The Ethiopian approach so far has been technocratic and
hierarchical. There was no farmers' participation in the design
of conservation techniques, and there is little or no participation
in the planning of measures for a particular area. The rules –
reforest slopes above 35 per cent, terrace slopes above 20 per
cent, treat gullies – allow little variation to suit local preferences.

Perhaps the most serious shortcoming was that the Ethiopian
government made food aid the core of its conservation programme and food aid was limited. Despite the vast area covered,
and the massive armies of workers mobilized, the project came
nowhere near the scale of effort demanded by Ethiopia's crisis.
The project catchments were chosen from the most seriously
degraded land in Harerghe and along the main road north from
Addis Ababa to Asmara. More inaccessible and less seriously
affected areas were left to continued degradation. After ten years
– five of those at full steam – only 4 per cent of the land suffering
from serious degradation had been treated. By 1985, three times
as much cropland degraded to poor pasture each year as was
reclaimed by the programme. Even if the programme expanded
at 20 per cent a year, it would take 50 years to cover the areas
susceptible to erosion. Many of them would be eroded beyond
rescue long before then.

In the light of all this the Ethiopian Highlands Reclamation
Study recommended in 1985 that conservation should no longer
be considered a separate activity: it should become the basic
foundation of development. Conservation should be fully integrated with agricultural work. Conserving methods of soil and
crop management should be taught, alongside methods of intensified production. In early 1986 the signs were that the Ethiopian
government had realized the shortcomings of the existing programme and accepted the need to shift into a higher gear.
Preparations were being made for a conservation programme
covering the whole country. The energetic head of the Ministry
of Agriculture's conservation department, Kebede Tatu, told me
that every farmer would be expected and assisted to protect his
own fields, and every Ethiopian, young or old, would be expected

to plant three to five trees per year. All urban communities and all peasant associations would be expected to plant enough trees to meet their own requirements for fuelwood and poles. These activities would be backed up by grassroots development workers and foresters, with a massive programme of training of peasant conservation cadres paid for by each community. The food-for-work programme would become, in retrospect, the school and testing-ground for a programme on a scale equal to the measure of the challenge.

MINIMIZE COSTS, MAXIMIZE BENEFITS

Two central ingredients are needed for any successful soil conservation programme. One is a set of simple, cheap and effective techniques that enable farmers themselves to become the chief agents of soil conservation. The second is a channel for disseminating those techniques.

The choice of methods is crucial. In most programmes this has been seen as a technical problem. Terracing, the standard solution, has serious drawbacks. Machinery should not be used because of its unreliability and lack of peasant involvement, but without machinery terracing is very labour-intensive. Bench terracing in Sierra Leone, for example, takes around 100 man-days of eight hours for every 100 metres – 25 times as much as stone lines or trash lines. In Ethiopia, where terraces are not levelled off, fifteen man-days are needed for each 100 metres of hillside terracing, but even this is ten times as much as planting tree seedlings at ten to the metre. Terracing also takes up land previously used for food production: the steeper the slope, the more land it takes up, and the longer it takes for unlevelled terraces to level themselves off. In general, terracing should be a method of last resort.

Future soil conservation programmes should choose their techniques with three central economic objectives in mind. *They should cut down to a mimimum any land lost for food production. They should be as low-cost as possible, not just in capital and cash, but also in labour input. And they should make every effort to compensate for land and labour costs by increased production of food, fodder, fruit or fuelwood.*

Infiltration zones are bands of fodder grasses or tree crops with trash on the ground. They can cut down erosion and run-off, gathering water for the crops below, without diverting any land or labour from production

Stone lines (p. 165) and trash lines are among the lowest-cost methods in terms of labour. Another is grassed strips, 3–8 metres wide, aligned along the contour, which slow rainwater run-off and allow it to infiltrate. If the strips are planted with perennial fodder crops, and thatching grass, the 10 per cent or so of land lost will still be usefully productive. The same double purpose of protection plus production is served by infiltration zones. These are bands about ten metres deep, running across the slope, and planted with woodlots, or bananas and coffee with a trash layer below. Such zones trap soil and rain, which helps water cereal crops planted lower down the slope. Windbreaks (p. 179) can be planted with multi-use species so they not only protect the soil against wind erosion, but also produce a range of useful products.

The most promising method for moderate slopes may be to use very closely planted hedges of leguminous trees and shrubs like *Leucaena* or *Sesbania sesban* along the contours, every 5–10 metres. Once the seedlings are established, brushwood and other trash can be laid horizontally on the uphill side. The hedgerows will then act as a soil stop. Over a number of seasons rain will wash soil from the back of each plot to the front, levelling out the terrace automatically. The hedgerow can then be managed as in alley cropping (p. 192) and will produce protein-rich fodder, nitrogen-rich mulch, and fuelwood.

Many of the simplest and cheapest conservation methods are not separate specialized activities. Crops and soil can be managed during food production in ways that reduce erosion, either by increasing the protective cover of vegetation, by improving the structure of the soil with extra organic matter, or by slowing the speed and volume of rainwater that runs off, carrying soil with it.

Intercropping (p. 108) increases the amount of ground cover and the length of time it is in place, especially when one of the crops is low-spreading, like cowpeas or groundnuts. Crop-breeders can select varieties for their protective value. Fast emergence of spreading first leaves can establish an early soil cover: first rains, falling when the soil is bare, are usually the most erosive.

Many African farmers in dry areas use a wide variety of furrows and depressions in the soil around crops, mainly to trap water – but they also trap eroded soil. In wetter areas they plant on ridges and mounds to improve drainage but the hollows in between also trap soil and water.

Mulching (p. 107) achieves several anti-erosion effects at once. It provides a dense cover for the soil, increases its organic content, and provides a rough surface that allows rain to filter in more gradually. On a 10 per cent slope under maize, mulch reduced run-off from 42 per cent of rainfall to only 6 per cent. Soil losses were cut even more dramatically by 97 per cent – from 27 tonnes per hectare to less than 0.8 tonnes.

In semi-arid areas, ploughing helps with soil conservation by breaking up any crust that might be keeping water out, and roughening the soil surface. A chisel plough, with a narrow bent chisel instead of the usual ploughshare, involves less labour and less disturbance to the soil.

FARMING WITHOUT THE PLOUGH

In the more humid areas, however, there is strong evidence that ploughing can speed up erosion and soil degradation. Much of this evidence has been accumulated by Dr Rattan Lal, soil scientist at the International Institute of Tropical Agriculture (IITA) at Ibadan, Nigeria. When Lal arrived at IITA in 1970 he subscribed to the conventional wisdom that deep ploughing was essential if crop yields were to improve. To prove the point, he started two experimental plots on a medium slope at IITA. One was ploughed by tractor, the other was farmed without ploughing.

The results were the opposite of what Lal expected. The deep-ploughed plot yielded 5 per cent more than the unploughed plot in the first year, but lost 25 tonnes of topsoil per hectare. At first Lal thought the losses were a temporary effect of land clearance, but they continued, year after year, and yields declined. One part of the ploughed plot had been originally cleared by bulldozer and root-rake. This section, typical of much large-scale commercial farming in Nigeria, is now an object lesson on how not to manage soils in humid areas. In its lower reaches, where run-off accumulates and accelerates, there are already 30cm-deep gullies. In places, 15–30cm of topsoil have washed away, baring a layer of iron and aluminium salts which has hardened to rock. 1985 was a wet year: run-off at the base of the slope was three feet deep at times, and so powerful that it pushed down the cement wall that funnels water into Lal's measuring devices.

The neighbouring plot has been farmed without ploughing, using no-till methods. This involves killing weeds with herbicide, and inserting seeds through the killed sod. No-till farming offers very important benefits in humid areas, in reduced erosion and increased crop production. The mat of dead vegetation acts like a mulch and protects the soil against the rain. Run-off is cut by two thirds. More water filters into the soil for crops to use, and soil loss is cut by as much as 99 per cent. The whole environment for crop growth is improved: soil temperatures and acidity are reduced; the content of the soil in organic matter and nutrients is raised; earthworms and ants thrive. Their channels aerate the soil and allow plant roots to grow deeper and denser – another protection against dry spells. 'There is no real need for ploughing,' says Lal. 'Earthworms and ants become your ploughs.' As a result, crop yields are higher and more stable. In one comparison over twelve consecutive seasons the *tilled* plot produced 2.7 tonnes of maize in two seasons in the first year, falling off to only 1 tonne in the sixth year, despite the fact that it was terraced.

The *untilled* plot was not terraced, but produced 2.8 tonnes in the first year, and 3 tonnes in the sixth year. Fertilizer was used on both plots, but Lal believes that minimum tillage could allow more or less continuous cropping of the same land without fertilizer, if a vigorous legume like *mucuna* was grown every third or fourth year to fix nitrogen and provide a mass of organic matter. This system could give stable maize yields of around 1 tonne per hectare. This is more than the Nigerian average, which is achieved with extensive fallowing and soil mining.

Minimum tillage is *environmentally* appropriate for most of the humid tropics, though not for the crust-forming soils of semi-arid areas. It also involves much-reduced labour requirements, as it cuts the need for ploughing, weeding or terracing. The main problem is that it involves spraying with weed-killing chemicals. These cost money which poorer farmers do not have. The chemicals and sprayers have to be imported, and are therefore vulnerable to foreign exchange crises and import restrictions. However, in countries like Nigeria or Ivory Coast, where extra workers have to be hired at peak times, the saving in labour costs may more than outweigh the cash cost of the chemicals and sprayers.

Most humid-zone soils should not be ploughed. So far, the tsetse fly, by banning cattle from most of the zone, has prevented ploughing. Better-off farmers are often overcoming the *labour* problem with tractor ploughing, only to create a much more serious *erosion* problem. Tractor ploughing should in future be cut to a minimum, and replaced with minimum tillage. Among smaller, poorer farmers, minimum tillage will take much longer to spread, but in the long run it seems likely to be the most important soil management technique for the humid zone.

LINKING CONSERVATION WITH PRODUCTION

Devising techniques is never enough: they have to be disseminated as widely as possible. *Ease of dissemination has to be an important factor in the choice of method.* Conservation techniques must work on the farm, as well as the research station, and the only way of making sure that they do is to design them in cooperation with farmers in the major regions of each country.

Techniques are most likely to be suitable, and to spread quickly, if they are based on improvements of familiar traditional practices. They have to be simple to master, so that farmers can do them for themselves without waiting for experts or expensive equipment to come along.

The best channel for dissemination is not a specialized service but, as in Kenya, a nationwide network of all-purpose agricultural extension workers. Conservation work should be fully integrated with other aspects of agriculture and tree-planting on farms. One of the reasons that soil conservation has not so far made much headway in Africa is that it has been designed solely with conservation in mind. The major benefits usually envisaged are long-term: avoiding the gradual decline in yields that would otherwise occur. African farmers are no less prudent than any others, but the hazards of their environment make this year and next year their overwhelming concern. They cannot think of preserving the soil for future generations if that interferes with the task of feeding the present generation.

Yet there need be no conflict. In the African environment all methods of soil conservation increase the amount of water that filters into the soil for plants to use, rather than washing away. *Hence soil conservation is also water conservation.* Improved water availability provides better protection against short spells of drought in the growing season. Because organic matter and nutrients (including fertilizer) are richest in the uppermost layers of soil that are the first to wash or blow away, *soil conservation is also conservation of nutrients.* Because it improves water and nutrient availability, *soil conservation increases crop yields everywhere, in most cases from the first year,* with more pronounced benefits in the drier areas.

Appropriate methods of soil conservation can have dramatic effects on food output. Yield increases of 30–65 per cent have been reported from terracing, of 20–160 per cent from intercropping, of 200 per cent from minimum tillage, of 50–60 per cent from mulching, of 20 per cent or more from shelter belts. The FAO study on population-carrying capacity found that soil conservation measures alone could increase potential food output in Africa by one third. If methods that produced additional fodder and fuel were used – like alley hedges, grass strips, planted earth bunds – then the total increase in useful output could be very dramatic.

Using these approaches, soil conservation, and its hand-maidens, water conservation and agroforestry, offer the most affordable and acceptable approaches to boosting food production that are realistic in Africa's present circumstances. They offer a sustainable hope for the future, as well as an immediate improvement for the present.

8 Fighting the desert: land reclamation

Desertification is the most dramatic and alarming form of land degradation: the total loss of the biological potential of the land, the reduction of once-fertile earth to desert-like conditions. It is the final stage of all the processes of erosion, overcultivation, overgrazing and deforestation, the stark expression of the African environment's extreme sensitivity to abuse.

When I visited the Yatenga plateau in Burkina Faso in 1976, desertification was only just beginning: you had to look hard to find bare patches amid the cultivated fields and fallow bush. When I went back in 1985, they were everywhere. The rainfall that year was good, and at the time of my visit, in November, the bush should have been knee-high in waving yellow grasses. Instead, vast expanses were totally bare of any ground-level vegetation, covered with a fine, hard crust where silt clogged the soil's pores and kept the rain out, or with layers of rusty red gravel, all that was left of the topsoil. In some places the shrubs were still alive, in others they had died back to brittle frames, many snapped off at the base for fuel. The trees, with their deep roots reaching down to the water table, held on longest, but in places they too had died, and had been severed back to stumps too thick for local axes to sever. It is estimated that 15 per cent of the total area of the Yatenga has been lost under these bare patches. Many villages have lost a third to a half of their cultivable land, virtually all the plateaux and upper slopes, leaving only the flatter valley bottom land still capable of producing a crop. And now that the rain no longer soaks into the uplands, but washes off in surging sheets and silt-laden torrents, even the valley lands are threatened.

The greatest impact of desertification lies not in any inexorable advance of the frontiers of the Sahara – for its fringes support only the sparsest of nomad populations – but in such spreading blots of barrenness in agricultural land. The desert is infiltrating pasture land and cropland from within. Desertification is self-propagating: as expanding areas become useless for crops or livestock, the pressure on the islands of remaining fertility increases. Nomads take to farming, often beyond the limits of sustainable rain-fed agriculture. Whole families, sometimes whole clans or villages, migrate to better-watered areas. There they begin the process of deforestation, overcultivation and overgrazing anew.

The dangers are spreading (map, p. 353). A recent survey of desertification trends estimated that as much as 742 million hectares in Africa – 26 per cent of the total land area – was undergoing moderate or severe desertification. This represents no less than 85 per cent of the dryland area in Africa – a larger proportion than any other region in the world. A total of 108 million people lived in the affected zones in 1983 – of those, 61 million were in areas undergoing severe desertification.

The most serious threat hangs over the Sahel. A 1984 study by US geographer Leonard Berry found that the problem was getting worse in fifteen of the nineteen countries in the region. Berry scored each country on five aspects of land degradation – sand dune encroachment, rangeland deterioration, forest depletion, degradation of irrigation systems, and problems with rain-fed agriculture – on a score of one for some increase, and two for significant increase. Adding up the scores for the five aspects, twelve countries scored five or more out of a possible total of ten: these were Burkina Faso, Sudan, Djibouti, Cape Verde (5 each), Somalia (6), Gambia, Senegal, Mauritania, Mali, Niger and Ethiopia (7 each), and Chad (9).

Until very recently, neither donors nor affected governments have given the fight against desertification the priority it deserves. Less than 3.5 per cent of the aid to Sahelian countries in 1980 was directed at desertification control.

HALTING THE MOVING SANDS

But now a decade and a half of low rainfall, punctuated by two severe droughts, has shaken many Sahelian governments into action. Mauritania, Mali, Burkina Faso, Niger, Chad, and

Ethiopia have all made the war against the desert a central priority.

In the last chapter we looked at ways of protecting cropland that was still producing food. Rescuing land that appears totally lost to production might seem an impossible task. But low-cost, quick-acting techniques are now being developed to do the job. They are, quite literally, rolling back the desert's advances. They are symbolically the most heartening of all Africa's breakthroughs. True, there are some forms of desertification which are effectively irreversible: nothing can be done with soil eroded down to bedrock, or with laterite that has hardened to stone. Fortunately, these extreme forms of desertification are, as yet, very limited in extent.

Saline irrigation land can be reclaimed by planting salt-tolerant trees such as eucalyptus, *Prosopis juliflora*, or *Acacia nilotica*. These species reduce waterlogging and open up the soil so salts can be leached down from the surface. Deep gullies can be stabilized and used to grow fodder and useful trees. The crusted wastes of the Yatenga, as we shall see, are being reclaimed by simple lines of stones (p. 165). The barren rings that have developed in rangelands, around boreholes and relief camps, can come to life again spontaneously if animals are excluded for a year or two. Even the desert sands themselves can be reclaimed.

From November to February in the West African Sahel, the hot, dry Harmattan wind blows, veiling the landscape in dust. On some days it is as thick as fog and air flights are grounded. A large part of the soils of the Sahel are wind-blown deposits of sand, with only a bare minimum – as low as 2 per cent – of clay. Sandy soils are less easily eroded by rainfall, and do not so readily form a crusted surface, but they have balancing disadvantages. They do not hold water well. When the wind speed is high enough, and the vegetation that normally holds the soil in place is sparse enough, sand takes to the air, and the earth begins to move. The earliest rains of the season are preceded by wind storms carrying clouds of dust and sand. Sand collects against the smallest obstruction, burying millet seeds too deep for them to emerge, building little mounds over young seedlings. Farmers plant up to forty seeds in each hole, to make sure four or five survive. In a bad year they have to replant again, two, three or four times over.

Sand piles against the mud walls of compounds and topples them over when they are weak. It heaps around the stems of shrubs. A line of bushes can give birth to a baby dune. The dune grows, perpendicular to the direction of the wind. Along its sides a thin skin of blown sand hovers; at its crest a plume of sand streams out horizontally and curls down. The dune moves.

At Assoro, in the Bouza department of Niger, a monstrous dune, two kilometres long, accumulated by the banks of a seasonal river. Like the back of a half-buried brontosaurus, it loomed over the gardens in the river's flood plain, where sugar cane, sweet potatoes and cassava grew in the dry season from the residual moisture. In the rains, segments of the dune collapsed into the river bed. In the dry season, the dune was a source of sand, feeding other dunes to the south and east. Standing on its peak, you could see smaller dunes, three to five metres high, 50 to 100 metres long, scattered every kilometre or so across the countryside. Not only were the dunes themselves lost for agriculture, the shifting sands they harboured posed a continual threat to the millet crop.

Muhammadu Salifou, chief of Assoro, can still remember a time when there were no dunes. When he was a child, in the late 1940s and 1950s, the area was extensively forested. Gradually the forests were cleared for cropland. Trees were felled for fuelwood, and many died in the drought of 1973–4, along with most of the animals. 'When there were trees, we burned wood,' Salifou explains. 'Then, when there were still a lot of animals, we burned dung. When the animals died, we burned millet stalks.' The soil was progressively stripped of cover and of organic content. Crop yields declined and the domestic energy crisis intensified. Parallel with these developments, and partly caused by them, the dunes began to accumulate, from the latter half of the 1960s. The local villagers had no traditional techniques to combat the problem; their response was the progressive abandonment of cropland. By the late 1970s, the dunes had become a serious threat not only to the prospects of the village, but even to its existence.

In 1978 the local forestry department, with the support of the

American development agency CARE International, began the work of fixing the dunes. Three tree nurseries were set up and staffed, growing seedlings of eucalyptus, *Prosopis and Acacia*. Long pallisades of millet stalks were erected, one and a half metres tall, their feet buried 30–40 centimetres into the sand, to act as windbreaks to protect the young trees. Then the seedlings were planted at the beginning of the rainy season, so they would need no watering to get established. The villagers provided free labour, making the fences and digging the holes for the seedlings. Up to 80 villages at a time turned out for as many as 76 days in a year. They also provided the millet stalks, but as these represented a real sacrifice of fuel or fodder for livestock, CARE paid them 50 francs CFA (about 10 US cents) a bundle, enough to fence one metre. The pallisades were spaced ten metres apart: for every hectare of fixed dune, one kilometre of millet windbreak was put in. Roaming village goats were a real threat to the saplings. Barbed wire fencing was too expensive, so village guards were appointed and paid $50 a month, to protect the newly planted areas.

One in five of the trees did not survive. Mortality was specially high among the *Tamarix* and *Parkinsonia*. But the eucalyptus did well, putting on 1–2 metres a year. By 1982, when CARE's involvement in the project ended, no less than 22 dunes had been fixed with 45,000 trees, and 50 kilometres of millet pallisades had been erected.

By 1985, when I visited the area, the eucalyptus on the great dune of Assoro were 5–7 metres high, their thin, silvery leaves rustling in the breeze. Their neat rows pinned down the sleeping dinosaur as effectively as the Lilliputians had immobilized Gulliver. The millet pallisades had rotted, providing organic matter for the soil.

CARE continued to finance the central nursery, while the Niger forestry department took over the other costs. In the nursery, piles of sand and compost were ready to fill the black polythene planting tubes, and neem and eucalyptus seeds were beginning to germinate, ready for planting in the following June. Seedlings planted on the dunes in June 1985 were already a healthy 2 metres high by November.

The nurseries continue to produce a surplus of seedlings,

beyond the needs of dune fixation. These are given out free to farmers, to plant in their compounds and their fields. In the first year, far-sighted Muhammadu Salifou planted a small woodlot of eucalyptus. Dozens of other farmers have followed suit. In 1985 Salifou cut the first harvest, six 15cm-thick poles, 3 metres long, ideal for major roof beams. He was loading them onto a grumbling camel when we arrived.

The project was an indisputable success, despite the fact that it was a top-down affair, designed by experts and officials without consulting villagers. They were simply informed of the plans, and told of the benefits, in 'animation' sessions. Participation was limited to the provision of free labour. But the enterprise did respond to a profound felt need of local people, indeed an imperative of survival. The costs imposed on them were small: most of the work was done in the dry season, when other demands on their time were low. The land involved was useless for crop production. The benefit was clear and immediate: a reduction in the threat to their lands. They earned, too, from the sale of their millet stalks, and by the employment of locals as guards and labourers in the nurseries. The very favourable benefit – cost ratio, and the speed of pay off – made up for the initial lack of participation.

There is no doubt that a stronger element of popular participation and local initiative would put the programme on a firmer footing. There is still too much dependence on government action. I asked Muhammadu Salifou if the villagers would continue the work if the forestry department pulled out. He said they couldn't, because they were only poor people: he imagined the villagers having to pay the salaries of guardians and nursery staff.

Yet there was nothing in the project that the villagers could not have done for themselves with training: a community nursery, tree seeds collected locally, collective work in building the pallisades and planting, a rota of children for guarding, provision of millet stalks without payment. Careful and controlled cutting and replanting of the earliest dune plantations would allow the lost land to provide some benefits, not simply protection against losses. The nearby villagers of Yegalalan have found an even cheaper approach by using the succulent euphorbia to fix dunes.

It can be propagated simply by sticking cuttings into the ground. It is unpalatable to goats, so there is no need for guards. With reduced costs, and greater local responsibility, the programme of dune fixation could move ahead much faster and become invulnerable to fluctuations in government budgets.

REGENERATING THE NATURAL FOREST

At Assoro, and in the Yatenga, it is once-fertile farmland that is being rescued or defended. But bush and forest are just as seriously threatened by desertification.

The Guesselbodi forest, 25 kilometres east of the Niger capital, Niamey, is in an advanced state of degradation. It was gazetted for state protection, in 1948. At that time it was rich, dense woodland, dominated by shrubs of *Combretum* species, excellent for fuelwood, and at ground level by *Zornia*, a nitrogen-fixing herb well liked by livestock. The forest thickets harboured a varied wildlife. Local villagers remember giraffes, hyenas, jackals, warthogs, monkeys and buffaloes.

Official protection was to no avail. Colonial armies cut wood from the forest. A few bold farmers cleared patches for cultivation. Probably the most serious threat came from the herds of goats, sheep, cattle and camels, which foraged in the clearings between thickets of bush. Their numbers grew in line with human numbers – but the carrying capacity of the forest did not increase. Every year the protective ground cover of vegetation grew thinner. The clattering of hooves and the pattering of raindrops compacted the earth, blocking the larger pores with fine particles and forming a crust that the rainwater sheeted over without sinking in. Over large areas no ground vegetation at all grew, and even the shrub layer, deprived of water, died. Two severe droughts, in 1973–4 and 1984, killed off many of the trees.

Aerial photos taken in 1979, compared with shots from 1950, show that between 40 and 60 per cent of the cover disappeared over those three decades. The dark, dense vegetation became patched with bare expanses of white, and round the fringe of the plateaux deep gullies formed. By 1983 most of the topsoil had washed away. Rainwater that once filtered gently into the forest soil, replenishing the water table, now washed down through croplands, spreading the damage of erosion.

Villagers could, in theory, have got together when things began to go wrong to reverse the process. In practice they had no power to control use of the area. The forest was state-owned, controlled by government foresters who had in theory to be approached for permission even to lop a single branch of a tree. Villages could control their own livestock numbers – but not those of other villages, even less those of the nomadic Tuareg, Fulani and Bouzou herders who grazed the forest in the driest months. Local farmers did not possess the techniques to rehabilitate the forest once it was degraded – nor, even if they had, would it have been worth their while to do so, so long as the pattern of control remained unchanged.

In 1981 the USAID-financed Forestry and Land Use Planning (FLUP) project took on the challenge of the Guesselbodi forest. All but a handful of previous forestry projects in Niger had been large-scale plantations or village woodlots, created either by bulldozing existing brush and replacing it with untried exotics, or by seizing good cropland, without compensation. Not surprisingly, neither approach had much success.

The FLUP project tried a novel angle. The natural forest, adapted as it was to local ecological conditions, was not as unproductive and useless as Western experts on lightning appraisal missions had assessed it. Rough guesstimates had suggested that one hectare of natural Sahelian forest could yield no more than 0.5 stacked cubic metres of wood each year. More thorough study showed that the productivity could be two to four times greater than that. The economics of exotic plantations were based on false premises. It could be far cheaper, and more profitable, to manage the natural forest than to uproot it and replace it with eucalyptus or neem plantations that cost between $700 and $5,000 per hectare to establish. Because of these high costs, the rate of plantation in Niger had never exceeded 7,500 hectares a year and could not hope to catch up with the estimated annual deforestation of 60,000 hectares or more.

The FLUP project set out to develop a master plan to manage Niger's natural forests. The first step was to find out just how much forest there was and what species it contained, from

landsat imagery and aerial photos, calibrated by sample checks
of vegetation on the ground. Then a thorough survey was made
of the wood market in Niger. Finally the project would develop
techniques for sustained management of the natural forest based
on research into its ecology and growth rates.

The practical research was to be undertaken at a number of
model sites. Guesselbodi was chosen as the first of these. From
1981, detailed surveys were made of slopes, soils and vegetation,
and of past and present uses of the forest by local villagers, and
their preferences for future development. By November 1983 a
management plan had been devised, and was presented to the
villagers. 'At first, there was no response,' said project adviser
John Heermans. 'Then one guy got up and said, "We don't care
what you do so long as you hire our sons and pay them."' The
villagers had been consulted. They did not participate, but at
least they were willing to accept the project. It was dealing with
degraded land that did not belong to them. It provided jobs,
about seventy in all, filled from among the villagers. They stood
to lose little or nothing, they gained some immediate benefits,
and there could be long-term pay-offs.

The forest was divided into ten 500-hectare parcels, with a
view to a ten-year rotation. The first and second parcels were
badly degraded. Heermans and his colleagues looked for simple,
small-scale, low-cost rehabilitation measures which would provide immediate results. Some of them involved water-harvesting
techniques pioneered in the Sahel by Oxfam (p. 165): earth and
stone banks, aligned with the contours to slow down the flow of
sheeting rainwater, and dam it back up the slope for slow
infiltration; in the gullies, check dams of large rocks and intercep-
tion dams to divert the water of flash floods along gently sloping
dykes, onto the slopes; in between the banks and dammed
gullies, a patchwork of micro-catchments, small banks in the
shape of half-moons, Vs and Ys, which collect water running off
from uphill and concentrate it towards their low point. The
raised banks collect fine soil particles, twigs, leaves, and seeds.
Local species of trees, grasses and herbs plant themselves and
grow spontaneously in the moisture trapped by the banks. In
addition, project workers planted trees – mainly native species –

Water-harvesting: microbasins collect run-off from a wide surface and focus it on crops or trees

and valuable pasture plants such as the legume *Stylosanthes*, the tall grass *Andropogon*, and the vigorous shrub *Prosopis juliflora*.

Perhaps the most promising regeneration technique tested at Guesselbodi is the use of a mulch of twigs and small branches, too small for use or sale. These are scattered on the surface of a bare, deforested patch, with a few limbs pressed in to hold the rest in place. Like the earth and rock banks, the brushwood accumulates soil, sand, organic materials and seeds. It lowers the temperature of the soil beneath and protects it against battering raindrops. And it attracts termites from their craggy clay nests scattered through the forest. They build a layer of digested clay around each stem to protect themselves from the sun, and proceed to consume the wood. Termites are the earthworms of the semi-arid tropics. Their excreta are rich in organic matter, and they bore tunnels into the ground which open up pores in the soil crust, allowing water and air to penetrate and making root growth easier. To compare results, other strips were deep-ploughed by bulldozers, which opened up the soil crust to rain penetration and left a roughened surface with bumps and hollows that collected water. There were control plots of untreated bare land. In the first year, 1983, the control plots produced no vegetation at all. The plots mulched with twigs gave 210 kg of vegetation per hectare, dry weight. The ploughed plots yielded more than twice as much – 440 kg per hectare. But they did not keep their lead. 1984 was a drought year, and production on all plots was, on average, two thirds lower than the year before. By now the ploughed plots had begun to form a new crust, and production of dry matter was down 93 per cent, to only 30 kg per hectare. The mulched plots yielded five times as much – 150 kg, a drop of only 29 per cent on the previous year. Had 1984 been a normal year, they would have shown a healthy increase over the first year: the benefits of mulching with slow-decaying woody materials builds up over time. When I visited Guesselbodi in November 1985, two months after the end of the rains, the mulched strips were rippling with a dense growth of tall grass, herbs and self-planted shrubs, while growth on the bulldozer-ploughed strips beside them was thin and stunted. The results of mulch are heartening, says Heermans: 'They show that trees can be regenerated naturally by spreading a waste by-product that comes from exploiting trees.'

The newly treated parcels are protected against grazing animals for the first three years. Guards, mounted on camels, patrol the zone, impounding any animals they catch – though, as one guard lamented, catching an unco-operative cow single-handed can be a tricky job. The owners of impounded animals have to pay fines of $2.50 a head for cows and camels, $0.50 each for sheep and goats, stiff penalties for the occasional owner whose entire herd strays into the protected areas. If the animals are not claimed within 15 days, they are sold. The guards are Tuareg, to avoid any dangers of social pressures from local livestock owners. Locals were unhappy about the rules at first, but now accept them.

The ultimate aim of the Guesselbodi experiment is to devise a management plan for sustained yield of multiple products from the forest: fuelwood, poles, forage, honey, medicines and food. The first tests of cutting live *Combretum* trees showed a regrowth of up to three metres within a year. The *legal* cutting of live trees is a novel departure in Niger. It is normally forbidden, so there are vast stands of over-mature wood, currently wasted, that could be exploited for firewood. Agroforestry is another possibility. The project has handed over 30 hectares of land within the forest to local villages, on condition they plant and protect windbreaks and the leguminous tree *Acacia albida* (p. 75) in and between their fields. The micro-catchments, too, can be used for successful cropping of millet, sorghum and cowpea, at least for the first four or five years until the trees cast too much shade. Indeed the prospect of five years of crop harvests may well be enough to induce local farmers to build the water conservation works in the dry season, free of charge. Having seen the benefits in the forests, they may then be tempted to test them out on their own fields.

The economics of regenerating degraded forests look extremely promising. The project economist, Juan Sève, calculated the costs and benefits of making 400 micro-catchments on one hectare, planting 100 of these with cowpea and acacia, and 300 with millet and sorghum. The first-year investment would be around $400–$500 per hectare – considerably cheaper than the cheapest conventional plantations, which never start with totally degraded land. The income, from the second year, would be $142, rising to $195 in the fifteenth year as wood production built up. The annual rate of return would average a handsome 48 per cent – six to eight times better than that achieved by most Sahelian plantations. This calculation is based solely on the income from wood and crops. It does *not* count the benefits for neighbouring cropland of improved water infiltration, raised water tables, and reduced erosion, nor does it include the continuing value of the land which would otherwise be worse than useless. Four fifths of the first-year investment is in labour costs, at a relatively expensive $2.50 per man per day. This cash expense could be avoided if villagers built the earthworks in exchange for the crops that could be grown behind them. It is, in other words, a very conservative, conventional estimate of the potential benefits of rehabilitating degraded forest land. The economics of managing *undegraded* natural forest would be even more attractive. Sève is now working on an economic analysis that would include the costs of land degradation and the benefits of avoiding it. These factors are normally ignored, or assigned a very low value, in conventional cost-benefit analysis. 'There are millions of folks up here who are interested in going on living,' says Sève, 'and the wherewithal is disappearing. There has to be some economic value in that.'

The next step at Guesselbodi is to create co-operatives of local villagers who can eventually take over the running and exploitation of the forest. They will not have *full* control; the state will retain titular ownership, and forestry personnel will ensure that the rate of exploitation remains sustainable. But villagers will be responsible for day-to-day management, and they will share the benefits; it will be in their interests, and within their power, to make the forest a self-sustaining enterprise. The first co-operative was launched in 1986, covering nine villages, with a 45-strong council and a committee of nine. The co-operative was used as a channel for teaching agroforestry methods such as windbreaks and micro-catchments. Exploitation of firewood was scheduled to begin in the autumn of 1986. Heermans reports that the first meetings were very encouraging, and villagers came forward with information that was built into forest policy. Whether it succeeds or not, the Guesselbodi experiment has already shown that degraded forest can be quickly, cheaply and profitably rehabilitated. Even the wildlife is beginning to return to the protected patches. Hyena, gazelles and jackals have been sighted and I saw the footprints of ground squirrels and civet cat.

The desert need not win, nor (barring major climatic shifts) do we need any very elaborate or expensive techniques to push back its frontiers. Reclamation lies within the grasp of the peasants and herders of the semi-arid regions, and it can be made to pay.

9 Every drop counts: water conservation

The world weather machine has arranged things so that rain in most of Africa is delivered in inconvenient and potentially destructive ways. About one sixth of the continent is actually too wet, with rain for three quarters of the year or more. The high humidity fosters pests and diseases. Maize is attacked by armies of stem borers, and debilitated by leaf spot, leaf blight and leaf virus. Cassava falls prey to mosaic blight and white flies on its leaves, and nematode worms in its roots. Produce may have to be harvested wet and stored damp, so moulds and fungi can get to work.

Some 46 per cent of the continent is too dry, with less than 75 days' rain in the year – not long enough for even millet to mature. Another 8 per cent, with 75–120 days' rain each year, suffers from too variable rainfall. In these areas, swarms of quelea birds descend on crops from desert fringes, and the parasitic witchweed *Striga*, with its deceptively pretty purple flowers, blooms among the millet and sorghum stems and sucks nutrients from their roots.

Africa's rains are concentrated into one or two wet seasons. In the dry season, lasting up to nine months, no rain-fed crops can be grown. As we have seen, even the wet season is unreliable: the rains may arrive late, or end early, or pause unpredictably at crucial points in the crops' growth.

Even when it falls, rain is a mixed blessing to African farmers. It is a blessing when it infiltrates for crops to use – and a curse when it runs off, eroding sheets of soil uphill, gouging out gullies downhill.

Rainfall in temperate zones is, for the most part, regular and reliable. By and large it can be left to get on with the job. But where African farmers leave matters to nature, the result is permanent insecurity, the ever-present threat of famine, and the risk of erosion. Africa desperately needs to master water, to make up for the shortage or unreliability of rain with controlled supplies from dams or wells. It needs to master water to increase food production, harnessing the sunshine and labour wasted during the dead dry season, and providing additional security against drought. It needs to maximize the share of rainfall that soaks into the soil to water crops or replenish the water table, and to cut down the proportion that runs off uselessly and destructively.

Yet the control of water is less advanced than in any other developing region. One reason is that the African environment is less amenable to irrigation. There is less surface water and the evaporation rate is higher. Apart from the Zaire river, fed by the year-round rains of the Congo basin, African rivers are highly variable in their flow. In the dry season many Sahelian rivers dry up to their sandy beds. Of Burkina Faso's three major rivers, the Red, Black and White Voltas, only the Black Volta still flows in the dry months, slowed to a small stream – the only river in the whole country. Most underground aquifers are small, or very deep and expensive to tap.

In the African continent as a whole, some 9 million hectares were irrigated in 1985, about 5 per cent of the cultivated area. This compares with 8.5 per cent in Latin America and no less than 29 per cent in Asia. But almost half of all the irrigated area in Africa is concentrated in the five countries of North Africa. Of the 4.8 million irrigated hectares in Black Africa, well over half is in just two countries: Madagascar and Sudan, each of which irrigates about 15 per cent of its cultivated area. In the rest of Black Africa there are only 2.7 million hectares of irrigated land, a mere 2 per cent of the cultivated area. Much of this area has only partial control of water, and is therefore still vulnerable to fluctuations in rainfall: a 1979 survey by the Club du Sahel found that only one third of the irrigated area in seven Sahelian countries enjoyed total water control.

Traditional irrigation is little developed in Africa. A few tribes like the Chagga, the Sambaa and the Para of northern Tanzania developed their own elaborate networks of man-made channels (p. 78), but such active, complex systems are rare. Laborious irrigation works are not worthwhile where rain-fed land is plentiful. Many tribes use small areas that can be exploited without extensive works – valley bottoms with high water tables, river flood plains, even tidal swamps. The technology is basic – hand-dug wells lined with logs, small earth and stone dykes for control, and for lifting, buckets or *shadoofs* – buckets or gourds on seesaw poles, counterweighted by stones. These irrigated patches are viewed by farmers as part of complex farming systems rather than as the focus of intensive work: they provide a useful bonus in good years, a reserve against drought in bad. In many countries, particularly in West Africa, they have been expanding on farmers' own initiative. In Nigeria, irrigated valley-bottom land expanded from 120,000 hectares to 800,000 hectares in the twenty years up to 1978 – a growth rate of 10 per cent a year. This expansion has occurred with virtually no help from governments or major aid agencies.

Official investment in irrigation has been channelled almost entirely into ambitious and expensive large-scale projects, few of which have justified their costs. In the words of one senior World Bank agriculturalist, 'Irrigation in Africa has been a disaster area.' A major reason for failure has been the very high cost of formal projects – anything from $5,000 right up to $25,000 per hectare. Average costs are roughly double those of other continents. Because river flows vary enormously over the year, reservoirs must be built very large if they are not to dry up in the dry season. And most of Africa is relatively flat, so gravity is less useful for transporting water and pumping is needed. Africa lacks the basic infrastructure of roads, bridges and power supplies needed for large-scale construction works. These overheads, already present in many other areas, have to be added to project costs in Africa. Foreign contractors are normally used for construction. Tenders are bumped up to insure against the extra risks that African work is thought to involve – and often, too, to cover for 'commissions' for politicians.

With costs of $20,000 per hectare, the value of increased production would have to be $2,000 per year or more, or 5–6 tonnes of rice per hectare, to make the investment worthwhile. Few schemes reach such yields. The World Bank project at Lake Morondava in Madagascar, which cost almost $15,000 a hectare in 1981, achieved yields of only 2.6 tonnes per hectare. The massive Office du Niger area in Mali averages only 1.7–2.6 tonnes per hectare. Rice production from similar schemes often costs far more than imported rice to produce.

The low yields are not surprising. The large-scale schemes are usually designed by engineers, and based purely on physical considerations. Suitable crops are then chosen for promotion, with little or no regard for local needs or experience. Farmers are rarely consulted. Yet many of the big schemes presuppose a total change in almost every aspect of the farmer's practice: from extensive rain-fed farming to very intensive irrigated farming, often with unfamiliar crops, inputs and machinery. Pumping and control equipment is imported, and liable to the usual failures due to poor maintenance, lack of spare parts or fuel. The management of most projects has been highly centralized. Professionals ran the show and farmers, little better than labourers, were told what to grow and where. Low prices paid to farmers did not help to encourage efficient production.

The early years of the vast Rahad irrigation project in eastern Sudan was typical of the costly and damaging failures in this sphere. Covering an area about 25 kilometres wide and 160 kilometres long, the project cost a total of $400 million. The land required was taken over without compensation. The former owners became tenants, and outsiders were brought in, many of them nomads. Over the first two years of production, the yield of cotton, the main crop, dropped by more than a half. Those who paid the price of failure were the farmers. A USAID evaluation concluded that tenants with no source of income other than the official activities were worse off with the project than they had been before.

The basic flaw in the Rahad system was its centralization of all power in a bureaucratic corporation. The corporation took all decisions on crops to be grown, type and quantity of inputs, and the nature and timing of operations. It controlled directly no less than 80 per cent of the tenants' costs. Tenants who did not comply with corporation directives faced fines or eviction. The rationale was that experts must know better than illiterate farmers. The experience proved that the opposite is true. Corporation services such as tractor ploughing, seeding and weeding were often late, due to breakdowns, shortages of spare parts and fuel, lack of trained drivers and mechanics. Cotton prices turned out lower than projected, due to a depressed world market and artificially high Sudanese exchange rates.

As a result of the high cost and the experience of failure, progress in African irrigation has slowed radically. Between 1968 and 1974–6, the irrigated area expanded by 4 per cent a year. Since 1974–6, the annual rate of growth has fallen by more than half, to 1.7 per cent a year. Meanwhile older irrigation systems have deteriorated through lack of maintenance. In the Sahel alone, 26,000 hectares that were once developed are now no longer farmed. The expansion of irrigation in Africa has probably now reached a standstill, where new land coming under irrigation barely balances the losses of irrigated land through deterioration. Most major donors are now extremely cautious about investing in ambitious new irrigation projects in Africa. By 1985, the World Bank had virtually stopped lending for new large-scale projects.

GARDENING IN THE DRY MONTHS

In the longer run, Africa must make the most of her irrigation potential. The FAO has calculated that some 27 million hectares in black Africa could be irrigated with short transport of water – almost six times the present irrigated area. This potential, like everything else in Africa, is very unevenly and unfairly distributed. Almost 17 million hectares lie in four vast countries with no shortage of rain-fed land: Angola, Mozambique, Zambia and Zaire. At present these countries are irrigating less than 1 per cent of their potential. The Sahel, which needs it most desperately, has only 6.3 million irrigable hectares of which it is already using 2.4 million.

At higher levels of national income and development, conventional irrigation schemes may become possible. But over the next decade, the emphasis in irrigation must shift radically. The existing large-scale schemes will have to be rehabilitated, improving drainage and maintenance, decentralizing management, and giving farmers more say, and full autonomy in the choice of crops to be grown. Investment in rehabilitation is relatively modest, and can bring high returns.

New ventures must be more modest affairs that do not require large inputs of imported materials or machinery. As in all spheres, low cost must be a prime goal: when the investment is low, even quite modest gains can mean a worthwhile return. These will be projects that develop small-scale, localized sources of water – wells, streams, moist valley bottoms – rather than attempt to control vast catchment areas; projects that are largely operated and managed by local farmers, and that do not require sophisticated skills to maintain, that build on and improve what farmers are already doing, rather than trying to change it for unfamiliar, alien approaches.

One of the most seminal enterprises of this kind has been the series of dry-season gardening projects mounted in Niger by the US-based charity, Lutheran World Relief. Dry-season gardening is practised traditionally in low-lying areas in Niger, but not on a wide scale. The wells are hand-dug, shallow affairs, lined with logs. They last only a year or two before collapsing. Water output is sluggish, so even a small garden needs several wells, and the constant digging and collapsing of wells gradually degrades the land as topsoil is mixed with subsoil and the overall level gradually sinks. Constant supplies of thick logs are needed, and in many areas these are no longer available. Finally the gardens attract hungry livestock and have to be protected by fences cut from local thorn trees. The demands of fencing and well-lining add to the pace of deforestation. All these problems prevent the wider spread of well-irrigated farming.

Lutheran World Relief developed a very cheap technology that provided cement-lined wells for around $300 for a 6-metre-deep well. The wells are built up in one-metre sections, using a mould into which concrete is poured around steel reinforcing rods. Farmers themselves dig the hole, and provide sand, gravel, water and general labour. The well-linings are made by teams of local artisans, mostly farmers, trained by LWR. LWR provides the cement, the steel rods, and the teamwork, but the farmer pays back the cost over a three-year period. The money goes into a rolling fund which is used to pay for more wells. The rolling fund principle multiplies the impact of the initial investment. The projects train a proportion of farmers in gardening techniques, and they train their neighbours. Hand tools, improved seeds, fertilizers and pesticides are also supplied on a credit basis, fully repayable by the farmer. To protect the gardens, hedges of closely spaced *Prosopis juliflora* – a fast-growing, thorny legume – are planted. These permanent hedges not only reduce the need for wood fencing or expensive barbed wire; they can also be pruned or pollarded to provide good firewood.

Lutheran World Relief currently has a dozen gardening projects in Niger. Some of them have serious problems arising not so much from the basic model as from the administrative and social context of a country that is one of the most hierarchical and bureaucratic in Africa. Projects are administered locally by Niger government officials, who move on every two or three years. They are not always fully committed to project goals, or to the idea of co-operatives which increase farmers' autonomy and limit the power of civil servants. Some of the co-ops have been split apart by corruption among leaders. Members are afraid of funds being misappropriated, and will not make repayments or participate in group supply or marketing. Again there has been an over-emphasis on vegetables such as lettuce, onions or tomatoes; local markets cannot handle a sudden expansion of production and there is an unsaleable surplus. As one gardener put it, 'Lettuce does not fill your stomach.' There should be a much greater emphasis on staple foods – cereals like maize, rice and wheat, and root crops such as sweet potatoes or cassava.

Nevertheless, the projects have improved local incomes and the availability of food, and reduced the need for migration in search of work during the dry season. They have increased the rate of tree planting, and reduced soil degradation and the pressure on scarce wood supplies. They have worked a remarkable transformation in their local areas. Fields that were patches of dried grasses and sedum shrubs in the dry season are now lush gardens, thick and green with fruit trees and tall hedges. And they have had a strong influence at national level in Niger, providing a model for efforts on a far larger scale than a voluntary agency could afford. The Niger government, after the drought of 1983–4, made food self-sufficiency its top priority.

The chief means chosen was LWR's approach of dry-season gardening with living fences, operated by co-operatives.

SMALL-SCALE MEASURES ON A LARGE SCALE

Even the big agencies are now beginning to follow the lead of small voluntary agencies like Lutheran World Relief, adopting the small-scale approach in large-scale applications.

The World Bank-funded Kano Agricultural Development Project is the biggest of its kind in Africa, covering 800,000 farms on 26,000 square kilometres in northern Nigeria. The focus of the irrigation component is the 60,000 hectares of low-lying valley bottoms, or *fadama* land, in the state. Valley-bottom land usually has a water table close to the surface in the dry season, and fertile soils collected from the topsoil washed off uphill areas. Most *fadama* farmers already irrigate from shallow wells or from rivers, using the ancient *shadoof* – a large calabash on a pivoted pole, counterweighted by a stone. The *shadoof* is slow: even where enough land and water are available, a man can irrigate only a tenth of a hectare.

The Kano project's approach to irrigation is summed up by irrigation engineer M. Tarcisius: 'The emphasis is on appropriate informal irrigation schemes, using as far as possible traditional methods which are inexpensive, and where the benefits can be derived in a short period.'

The local Hausa farmers grow crops in small basins, often only two metres square. On uneven or sloping land the basins can easily be levelled to provide equal water for each plant and prevent nutrients from being washed away. The project's extension workers build on this traditional technique rather than trying to replace it. They recommend larger basins, four metres square, to cut the area lost to ridges by a half, while improved seeds, pesticides and fertilizers provide better yields.

The supply of water is improved by cheap shallow tube wells, 15–20 metres deep, costing only $900 each. Where water is flowing two or three metres below the surface in sandy river beds, wash bores can be used – tubular screens that let water through and keep the sand out. These cost only $150 each. To replace the laborious *shadoof* the project imported cheap Honda petrol pumps. The two-inch pumps cost only $500 and can irrigate a hectare and a half, fifteen times the area a *shadoof* can cover. The three-inch pump, costing $700, can handle three hectares. Local motorcycle mechanics were trained to service and repair the simple pumps. Farmers pay the full cost of the pumps, and half the cost of the wells or wash bores. The total cost per hectare under irrigation ranges from $500 to $1,250 – a fraction of the $19,000 per hectare price tag of Kano State's ambitious river basin schemes.

The returns to the farmer are extremely attractive; for an outlay of $500–$700 for the pump, plus another $75 or $450 if he needs a wash bore or tube well, he can expect an income of anything between $1,800 and $3,500 per hectare per year – a return of 60–500 per cent on his investment. 'I used to kill myself all day with the *shadoof*, and even then I could water only a quarter acre [0.1 hectare],' farmer Hussaini Chiroma told me. In 1984 he bought a three-inch diesel pump for $700, and irrigated one hectare. In the dry season he grew wheat, peppers, onions, tomatoes and sugar cane worth $4,000, plus half a ton of maize in the rainy season – return on investment, more than 400 per cent. The river he draws water from dries up in April and May each year, and food supplies run low. Chiroma's next move will be to request a wash bore, so he can grow food to fill the hungry gap.

The project has certain weaknesses. One is that the pumps programme is dependent on imports. These imports are guaranteed by the World Bank loan for the duration of the project, but when the project is over they may be vulnerable to foreign exchange crises. Even so, the import dependence is very much lower than for conventional irrigation schemes. Eventually pump manufacture would be a prime candidate for a Nigeria-based industry.

Another questionable feature is equity. The project makes it possible for individual farmers to appropriate the use of many times more *fadama* land than before. Only those with capital of $500 or more qualify. To some extent, the project is providing subsidies to make the rich richer. KNARDA encourages poorer farmers to form co-operative groups to buy wells and pumps. A credit programme – allowing farmers to repay well and pump costs out of their first year's profits – would open the programme more widely to the poor. And with the profits that beneficiaries can make, there is no justification for subsidizing the wells and wash bores; farmers should pay full cost.

Apart from that, the project's irrigation work is a notable success in an area where the World Bank has experienced many costly failures. The irrigation component has avoided high-cost ventures requiring highly skilled management. It has gone for cheap and simple improvements on existing practices. From the farmer's point of view, it does not depart radically from his existing experience, it is relatively low cost and pays for itself within a single season. The technology is easy to maintain in a society where small petrol engines and mechanics to repair them are increasingly common. And the scheme is self-sustaining, requiring a minimum of skilled management.

ADAPTING TO ARIDITY: THE NEW WATER ECONOMY

Water conservation is advisable in all but the wettest parts of Africa. In semi-arid regions, especially in the Sahel, it has become a matter of survival. The isohyets that map the limits of certain levels of rainfall have shifted south by 100–200 kilometres. Sahelian farmers find themselves caught with cultivation techniques and crop varieties suited to higher rainfall. They are adapting as best they can, picking out the seeds that do best in poor years, shifting from maize and sorghum to millet, which needs less rainfall, adopting techniques remembered from earlier droughts way back. But by and large they are not adapting fast enough.

A changed technology is needed; a body of techniques and crop varieties adapted to making the most of scarce water. We have seen some of the characteristics that crop varieties should possess in drier areas: early maturity, deep rooting, ability to withstand hot soil temperatures and sandblasting.

It is crucial in all but the most humid areas to increase the proportion of rainwater that filters into the soil. Infiltration not only feeds crop roots; it replenishes the soil's store of moisture, raises the level of the water table, and refills aquifers. Maximum infiltration is a protection against drought. Without it, any strategy based on exploiting underground water will be threatened as the water table gradually sinks, and wells have to be dug deeper and deeper, at higher and higher cost.

Technologies that are appropriate have been used in isolated pockets, by inventive farmers. But as a body they are unfamiliar and unconventional. They imply a mastery over slopes and surfaces and stones and water flows – a mastery that African farmers, with their experience in managing complexity, are well placed to achieve, provided they are taught in time.

One set of techniques for low rainfall areas is water harvesting – concentrating the rain falling on a wide area into a smaller area where crops are grown. This technique was originally developed in the arid Negev desert three or four thousand years ago; Negev farmers shaped hill slopes and surfaces so as to channel rainfall from large catchment areas onto their fields.

Waterspreading: floods racing uselessly or destructively down waterways can be checked, diverted onto fields, and spread by stepped basins

Water harvesting in Africa is more likely to take the form of micro-catchments. Rainfall from small areas of 16 up to 1,000 square metres is concentrated into confined basins or depressions where trees, forage and crops can be grown. Water harvesting is an excellent technique with soils that have formed an impermeable crust. The crusting, paradoxically, helps by increasing the amount of run-off from catchment areas that can be used to water crops in the basins.

Water spreading is the reverse of water harvesting[1]. A great deal of rainfall in semi-arid areas concentrates into rills and ravines where it is useless to crops. Water spreading techniques block the flow in these seasonal torrents with carefully designed rock dams. These divert the water along low dykes onto cropped fields, where it can infiltrate into the soil. A series of 'U'- and shallow 'V'-shaped terraces allow the water to spill gradually down the slope, like the layered basins in an ornamental fountain, with time to soak in.

Maximizing infiltration reduces the amount of rain running over soil surfaces and eroding them. As we have seen (p. 139) most soil conservation techniques also slow down run-off and increase the rate at which water filters into the soil. Mulching and minimum tillage, for example, cut the proportion of rainfall lost in run-off by as much as 85 per cent. Rough ploughing creates small depressions and crests that serve as micro-catchments in miniature, collecting rainwater and soil particles. In the Yatenga province of Burkina Faso, farmers plant their millet in shallow circular depressions called *zay*. These are holes 10–30cm in diameter and 5–15cm deep, spaced at distances of 40–70cm. The soil inside is loosened by hoe, and part is pulled forward onto the down-slope edge of the hole. Seeds are sown in the centre. The *zay* act as miniature windbreaks. They collect eroded soil and concentrate water around crop roots. Manure can be applied in the hole, where it will not be washed away. The *zay* also reduce the work of weeding, by giving the crops a head start over weeds growing outside the depressions.

Planting crops along low ridges is a widespread practice in Africa. In drier areas, they collect rainwater into the

[1] See p. 148 for detailed application of water harvesting and water spreading.

furrows where it can feed crop roots. Tied ridging, researched at the Semi-Arid Food Grain Research and Development (SAFEGRAD) station in Burkina Faso, links two such ridges with cross-ridges every 1.5 metres or so, creating rows of small rectangular basins around 20cm deep. Under research station conditions, tied ridging has increased maize yields by 1–1.5 tonnes per hectare, sorghum yields by 0.95 tonnes, and millet by 0.6 tonnes. When tested by peasant farmers, yield increases have been much more modest, but still worthwhile, especially on steeper slopes, and in drier years. In the drought year of 1984, maize grown between tied ridges yielded 360 kilos a hectare, while maize grown on the flat gave a mere 15 kilos.

PERFECTING A TRADITIONAL TECHNIQUE

The problem with tied ridges, holes and depressions is that they require greater labour, often at crucial times of the farming year. If this means that part of the crop is planted late, much of the potential gain could be lost. Cheap and simple improved tools or basic mechanization might help, but on the whole techniques that can be carried out with work in the dry season are more likely to be widely adopted.

One of the most effective approaches has been developed by Oxfam in the Yatenga area of Burkina Faso. It involves nothing more sophisticated than lines of stones ranged along the contour. At first sight it is hard to believe that these lines could make much difference. In fact, they increase infiltration, boost crop yields, reduce erosion, and are even capable of rehabilitating totally degraded land. The technique of making them is so cheap and simple that the stone lines are spreading with astonishing speed.

The Yatenga was once the prosperous heartland of the Mossi empire. It is still the most densely settled area of Burkina Faso, but almost two decades of poor rainfall, coupled with the increasing pressure of human and animal populations, have taken a devastating toll. Land that once yielded crops has turned to bare, hard plains of crusted sand or gravel where not a blade of grass grows, even at the height of the rainy season. After years of lowered rainfall and poor infiltration, the water table has dropped by up to five metres in places. Wells dry up sooner – at Rouko in the Kongoussi area, even 25-metre wells are empty from January to July, and women have to walk 10 kilometres to the nearest water source, where they face waits of up to five hours. To avoid the queues, many Rouko women set out at 5 P.M., and sleep all night at the well. They get up at 5 A.M. next day, arriving home by 7 A.M..

The Mossi developed their own traditional remedies to the water problem. Farmers in some areas place lines of stones, roughly along the contours, to slow down the destructive force of run-off. But slopes in the Yatenga are very slight, between 0.5 per cent and 2 per cent, and levels are impossible to get right by eye – sometimes it is even hard to tell if the ground is sloping up or down. If the stone lines do not follow the contour, their benefit is muted.

The Oxford-based charity Oxfam began work in the Yatenga in 1979. One of their field workers, former Peace Corps Volunteer Bill Hereford, paid a vacation visit to the Negev desert. He was deeply impressed by the water harvesting techniques he saw there and wanted to introduce them into Burkina Faso. The Yatenga project started life as an agroforestry venture. As so many tree-planting efforts had failed, the idea was to give the trees a head start by planting them in basins fed by micro-catchments. But the mortality rate of seedlings in the dry season was massive. Many villages did not have enough water to drink, let alone to water trees with.

The Yatenga farmers themselves forced a change in the project goals. Their absolute priority was food production. Project director Peter Wright worked together with villagers to perfect techniques for water conservation. It was a textbook example of barefoot science. Wright built on local techniques and farmers were his research collaborators, applying and developing the approaches that seemed most promising.

The most effective of all the techniques that Wright worked on was a development of the traditional stone lines. Wright found that if these are aligned properly with the contour levels, they hold rainwater back and make it pool for anything from four to fifteen metres uphill, giving it plenty of time to infiltrate into the soil. At the same time the lines are not completely impermeable.

Some water passes through the gaps in the stones onto the next levels, so the soil below the barrier does not dry out. The lines of *diguettes* (little dams) are easy to make. A trench, 5–10cm deep, is dug to provide a foundation, and the earth piled uphill. Then medium-sized stones, 5–15cm in diameter, are heaped up to a height of 15–25cm, in a band 20–30cm wide. The lower part of the wall on the uphill side is plastered with the excavated earth, to prevent run-off water from seeping underneath the stones and undermining them.

The effects of the stone lines are quite remarkable. They slow run-off and increase infiltration. The increased water availability and reduced leaching of nutrients lead to yield increases averaging over 50 per cent. The effect appears to be more pronounced in drier years. The lines reduce, and even reverse, the loss of soil. Sand, soil, leaves, twigs and seeds accumulate against the lines. Farmers in the Koumbri area did their own tests to confirm this, by pegging marked stakes into the ground. Where there were no stone lines, the soil level dropped by 10cm in a single year. Where there were lines, the soil level *rose* by 15–20cm. The lines are capable of turning most of Burkina Faso's barren expanses back into usable cropland.

Correct alignment of the lines is essential: they must run level with the contours. Originally levels were established using a quadrant, but this was an expensive and complex item. Wright's aim was to teach farmers to make their own stone lines, so it was essential to develop a cheap method which they could quickly learn to master. The answer was the hosepipe water level. Two stakes, about 1.5 metres tall, are marked half-way up with a series of lines one quarter of a centimetre apart. A narrow, transparent hosepipe, 10–20 metres long, is filled with water, and one end attached to each stake. Water is added or tipped out until the water level lies between the marked lines on the stake. Provided all air bubbles are expelled, whenever the bottoms of the two stakes are level, the water in each of the hoses should come up to the same mark on the stake, and the device can be used to mark out contour lines on a slope, hammering in pickets to show where the foundation trench should be dug. The hosepipe water level costs only $6 to make, and the technique is easily mastered. Illiterate peasants can pick

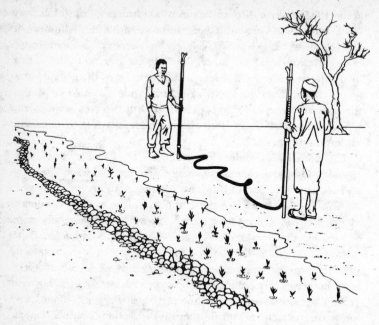

On gentle slopes, lines of stones, aligned with the contours by means of a cheap hosepipe level, back rainwater uphill where it can infiltrate slowly. The lines can rescue crusted wastelands

up the method thoroughly in a two-day course, and pass it on to their neighbours, or mark out the lines for them. Oxfam's practice-oriented training courses are held in villages that already have stone lines, so that novices can talk to farmers who already have the lines, and see the effect they have.

Motivation was hard at first. 'The people were very fatalistic,' the present project director, Mathieu Ouedraogo, remembers. 'They would say, "God has done this". We would say, "Do you remember when there were trees here, and the rains were good and the wells were full? And now the trees are gone and there is no rain and the wells are dry. Who cut down the trees? Who lit the bush fires? It was man who did it, not God, and man must reverse it."'

The stone line technology has spread rapidly, spurred on by drought and the deepening environmental crisis in the Yatenga.

In 1984–5 alone, Oxfam trained 600 farmers from 30 villages out of a total of 93 villages that asked for help. The techniques were also being disseminated widely by the Naam movement of popular co-operative self-help groups (p. 279).

The project's technology is not alien and unfamiliar: it is based on an improvement of traditional methods. It was developed in close consultation with farmers, ensuring that it took into account the realities of their situation. It is simple, easy to master, and extremely cheap, involving virtually no cost other than labour. The materials are readily available. The work can be done in the dry season, when it does not conflict with food production. The land actually taken up by the stone lines is only around 3–6 per cent of the surface area, and in any case most farmers first try out the technique on barren land. There is virtually nothing to lose. The yield increases are significant, right from the first year.

One of the most encouraging signs in the whole of Africa is the speed with which Burkina farmers are passing the technology on from neighbour to neighbour, from village to village, as they see the effects with their own eyes. In the Koumbri area, 20 villages with Naam groups have introduced the stone terraces. In December 1984, after a disastrous harvest, Saidou Porgo of Watinoma village covered four of his nine hectares of land with stone lines. A friend who had been trained marked out the contour levels for him. Porgo got the stones from hills three kilometres away, some by donkey cart, some on the family's heads. The rains in 1985 started well, but finished almost a month too soon. Fields with *diguettes* held a lot more residual moisture, and still gave a good yield. When I visited Porgo's farm in November 1985, harvesters were snapping huge heads from the millet and Porgo was stuffing them into over-brimming donkey carts. The heads were almost 40cm long and 3cm thick, full of fat, even grains. Porgo reaped 40 cartloads, almost three times his previous best, and to cap it all the water level in the well at the bottom of the field, which had been falling for five years, rose.

Saidou Porgo's example is having a visible effect on his neighbours. Bouréma Porgo's millet, in the adjoining field, had thin cobs, only 15cm long, with small, uneven grains. Bouréma intended to make *diguettes* that year. Another neighbour, two-metre-tall Moumouni Porgo, harvested only stalks for goats from his millet fields in 1984 and again in 1985. In the dry seasons he had to wander the Ivory Coast looking for paid work, buying millet for his family with his wages. 'This year,' he told me with grim determination, 'I will carry stones till I break my head to make the lines.'

Fatalism has been the common response to disaster in the Sahel. Now, perhaps for the first time, people are realizing that human beings can shape their environment, for good as well as for bad; people like Suleymane Ouedraogo, a tall, dignified Mossi from Gourga village. 'How is your country?' he greeted me, with a characteristic query. 'Is it all right? Are the rains good?' Gourga was one of the earliest places to try out the lines, and Ouedraogo one of the first volunteers in Gourga, back in 1981. 'I could see the land being eaten up day by day,' he explained. 'I had to do something. Some people said I was foolish, I was doing so much work for nothing, it was madness. I took no notice of them. When the rains came and I got better crops than they did, they saw that it was not madness.'

Ouedraogo's example was contagious. By 1983 most Gourga farmers had built stone lines. In the drought of 1984, those with lines harvested enough for three months, those without got nothing. The last stragglers were convinced: by the end of 1984, every single villager in Gourga had *diguettes* on some of their land, and the village development group built lines on some of the most degraded land in the area to make a collective field. When I passed through in November 1985 wind and rain had reconstituted the soil and the villagers were reaping a decent harvest of sorghum and groundnuts. 'We can't leave this land,' Ouedraogo explained. 'There are already too many of us going south. If you don't want to leave, you have to do something to save your land. The stone lines help us to get back the lost land for our children.'

10 Seeing the wood and the trees: forestry

The tropical forest is an exuberant explosion of life forms. It supports more species in one square kilometre than there are native species in the whole of Britain. It rains down three times as much weight in leaves as the temperate forest. The forest floor is layered with leaves and twigs, covered with earthworm casts, criss-crossed by ants' trails.

This extraordinary biological diversity and productivity is not the result of fertile land. The soils that support Africa's rainforests are often poor, thin, acidic, leached of nutrients, high in iron and aluminium salts. What makes them hospitable to life is the environment of the forest, incessantly recycling all available nutrients, increasing humidity, reducing soil temperatures and evaporation. *The forest creates the conditions for its own survival.*

Even outside the forests proper, trees are a key element in the stability of all but the most arid ecosystems in Africa. They are crucial in counteracting the effects of erosive, torrential rainstorms and infertile soils. Their roots anchor the earth. Their canopies protect it from the sweep of winds and the splash of raindrops. They draw up nutrients from deeper soil layers and deposit them by leaf fall on the surface, increasing the organic content of the soil, improving its fertility, its permeability for plant roots, its capacity to hold water. Trees increase the infiltration of rain into the soil, raising water tables and reducing destructive run-off.

Once tree cover is weakened or removed the benefits are curtailed. Soil temperatures rise to the point where plants and crops are stressed, and evaporation increases. Soil moisture, fertility and stability decline. More water is lost to run-off. Wind and water erosion increase.

Deforestation is therefore one of the gravest threats to ecological stability and food production in Africa. The level of that threat varies widely between regions. One third of Africa's land area is covered by forests as yet undisturbed by agriculture – a total of 703 million hectares in tropical Africa in 1980. Some 217 million hectares of this was closed forest, mostly in the Congo basin. Four fifths of the closed forest is undisturbed or unproductive, and likely to remain so for a long time to come; a mere 0.17 per cent is cleared each year. At the current rate it would take two centuries for this area to be reduced by one third.

The other one fifth is mainly logged over rather than properly managed. Trees of the most valuable 5 per cent of species are removed, leaving the rest in place. The remaining forest is often damaged by bruising or toppling of neighbouring trees, but the rainforest's power of regeneration, if left alone, is immense. The 20 per cent of closed forest that is managed or logged over is disappearing quite rapidly at more than 1 million hectares each year.

Outside the closed forests are the open woodlands, interspersed with grassland. These totalled some 486 million hectares in 1980, and were shrinking by 2.3 million hectares a year, or 0.48 per cent.

Regionally, the deforestation threat is most serious in the more densely populated areas of West, East and Southern Africa. East Africa's closed forests are being cleared at an annual rate of 0.8 per cent a year. Burundi's are declining by almost 3 per cent a year. Around 1940, 40 per cent of Ethiopia's land area was under forest; today the proportion is down to a mere 4 per cent. The fastest rate of deforestation is occurring in West Africa, where 4 per cent of the closed forest is being razed each year. In 1900 the Ivory Coast had 14.5 million hectares of closed forest. By 1980 these had been whittled down to 4 million hectares, and were declining by 300,000 hectares annually: at that rate they will have vanished by the mid-1990s.

All of these figures relate to the complete destruction of forests, when they are converted to cropland or grassland and subsequent bush or tree fallows. The main cause, responsible for at least two thirds of deforestation, is clearance for agriculture due to the pressure of population. Commercial forestry is the second largest culprit, opening up forest roads, clear-felling some areas. These operations also open up the forest to encroaching farmers. None of these figures include the progressive *degradation* of forest, woodland and wooded fallows, the gradual thinning out of cover. The major causes of degradation are threefold. First, repeated fires, deliberately lit by farmers to ease the task of clearing, and by pastoralists to force the desiccating grasses to sprout a few green shoots. Second, overcutting for fuelwood and charcoal. This usually starts with branches, gradually weakening the tree until it dies, when major limbs and trunk will be taken. Overcutting is a major problem around urban centres, along main roads that feed them, and in some semi-arid areas where fuelwood and charcoal are an important source of cash. Overgrazing by expanding livestock herds is the third main cause: removal of grass cover by cattle and sheep can lead to crusting (p. 140) so that even shrubs and trees dies. Goats in particular can weaken trees by eating their foliage, and make a quick snack out of tree seedlings. Overgrazing can seriously damage trees' capacity to regenerate themselves.

Deforestation often accelerates during a drought, when the trees' contribution in exploiting deep groundwater is most needed. Drought itself kills many trees, while humans under the compulsion of drought kill many more. Some pastoral nomads 'hinge' trees, cutting them half through at the base to make the tree push out a last dying flush of leaves for livestock to eat. And drought often forces people in stricken areas to destroy and sell off their tree stock to get cash to buy extra food. These processes leave areas more liable to drought, and more vulnerable when it does recur. Drought, too, creates the conditions for its own perpetuation.

A CATALOGUE OF FAILURES

Forestry efforts in Africa, to date, have fallen far short of making up for the losses caused by deforestation. African foresters have concentrated almost exclusively on conventional forestry, the management of forests and plantations to produce timber for industrial uses. They have been spectacularly unsuccessful. In

1960 around 4 million hectares of forest were intensively managed, but twenty years later the area had shrunk by more than half to 1.75 million hectares – less than 1 per cent of the area of closed forests. The remaining 701 million hectares of closed and open forests were left unmanaged and unprotected. The annual rate of new plantations in Africa, at 126,000 hectares a year, is little more than one thirtieth of the rate of deforestation – the worst ratio of any major region. In Latin America new plantations amount to one tenth of annual deforestation, and in Asia to just under a quarter.

Forest departments are too short-staffed and underfunded to do any more in the conventional way. In 60 developing countries surveyed by the World Bank, forestry budgets accounted for less than 2 per cent of combined spending on agriculture and energy. Burkina Faso had only one forestry field worker per 1,000 square kilometres in 1980. In one vast district of sixteen forest reserves, the eleven forestry staff had an annual operating budget of only $1,000, with a single vehicle between them, and a fuel allowance of 50 litres per month – enough for around ten kilometres a day. Burundi did not have a single trained national forester until 1980.

Of the local and aid funds that were spent, most were badly spent. According to US forestry expert Fred Weber, between 1975 and 1982 some $160 million was spent by public and private donors on forestry projects in the Sahel. The total area reforested was less than 25,000 hectares, and as much as one third of that area was producing very little wood at all.

Often, existing natural forest or bush, or even fertile cropland, was taken over for plantations and cleared with heavy earth-moving equipment. Exotic species were planted, with little or no experience of their performance in local conditions. Growth rates and wood yields achieved were often half the levels expected on the basis of performance in other continents – and far below what was needed to provide an economic return on the investment. Much of the wood produced was often unsaleable as fuelwood, either because the local market potential had been overestimated, or because consumers preferred native species. There is now some doubt whether plantation projects, with their high costs, can ever be viable in semi-arid areas, where trees grow too slowly to produce the required returns. The current feeling in major agencies such as USAID and the World Bank is that plantations should not be supported in areas with rainfall below 1,000mm.

When many plantations turned out to be expensive failures, village woodlots were tried. So far they have had little success in Africa. Often good cropland was taken over, local villagers did not always derive the benefits they were led to expect, and there were conflicts over responsibility for maintenance and the sharing out of work and of wood.

Both woodlot and plantation projects were planned by Western experts and government officers, and as usual local people were rarely consulted. Where villagers suffered no losses, they were at best indifferent to the success of projects, and their continued pressure to gather fuelwood and graze animals imposed high costs in terms of fencing or guards. Where they lost land, or customary rights to fuel, fodder and other products, they were often openly hostile and in some cases deliberately set fire to plantations.

Meanwhile, existing forestry laws worked against private or communal initiatives to plant trees. In most francophone African countries and in Ethiopia, trees are the property of the state. Where foresters are vigilant, as in Niger, villagers may have to ask permission to prune trees they planted themselves on their own land, and may be fined if they do not. Such regulations, designed to protect trees, have the opposite effect. They act as a powerful disincentive against tree planting, and a temptation to illicit felling in bush or forest areas.

MANAGING MOTHER NATURE

Of course plantations and woodlots will still be needed, especially to supply towns and cities. They *can* be made to work, in the right circumstances. In the case of woodlots local people would have to be much more closely involved in projects from the earliest phase of design, through implementation, evaluation and modification, to the enjoyment of final benefits. The location of projects should be fully agreed with local people, and farming, grazing or gathering areas should not be taken over without adequate compensation. The allocation of work and the distribution of benefits should be equitable, and clearly understood from the beginning.

Plantation projects are unlikely to be viable in areas with rainfall below 1000mm a year, because of slow growth rates. Elsewhere project organizers should be careful to consult local people on siting, compensating them fully for any losses. Preferably, people should be involved in some more direct way, such as being allowed to grow crops, in the early years in exchange for looking after the young seedlings – the *taungya* system, in use in parts of Ghana, Kenya and elsewhere – or through beekeeping or controlled grazing and gathering of fallen branches.

The most promising approaches for the future lie in two main directions. One is the management of natural forests. In 1985 there were no less than 685 million hectares of natural forests and woodlands in Africa plus another 178 million hectares of tree fallows, against a mere 2.4 million hectares of plantations. Most natural forests are severely neglected. Many are underexploited, with massive wasted resources; many are overexploited, fast on their way to becoming desert.

Natural forests are multi-purpose resources for local people. In Burkina Faso, for example, women collect the nuts of the shea butter tree to make cooking oil, while their children eat the furry, skittle-shaped fruits known as monkey bread from the great bulbous baobab trees. Families collect wild fruit and nuts, leaves and seeds for nutritious sauces, fibres for mats and ropes and baskets, home remedies from leaves, pods and roots, and chewing sticks to keep their teeth clean. In periods of drought the forest becomes an important reserve of famine foods and fodder. At all times it can be used for grazing, hunting, beekeeping, and fuelwood. And the potential productivity in wood alone is two to four times greater than some of the early studies suggest. All in all the total contribution to local welfare, income, employment and drought insurance is probably higher than those of most plantations of single-purpose exotics. The species are adapted to the local environment, and they regenerate themselves naturally as long as they are not abused.

The main problem with natural forest, especially in the semi-arid areas, is the danger of progressive degradation ending in desertification. Fires, deliberately started to clear bush, force grass to shoot, or drive game from its cover, are the main hazards. They destroy seedlings and coppice shoots and slow down the rate of natural regeneration. Overgrazing has a similar effect, though controlled grazing may be helpful in stopping the accumulation of inflammable dry grass – indeed, many tree seeds germinate faster if they pass through the gut of a sheep or goat and are deposited with a helpful dollop of manure. Controlled grazing and protection against fire alone can allow the forest to rehabilitate itself; fire-protected forest can yield up to two and a half times as much as forest that is exposed to burning.

In humid areas natural forests that are carelessly exploited by selective logging can be degraded to the point where they become a useless wilderness. In 1971, Ghana launched a project to rehabilitate her degraded natural forests. The technique was to fell whole areas, burn the debris on site, and replant with fast-growing exotics. The programme was quickly abandoned because of its high costs. Weeds invaded the replanted plots and smothered the tree seedlings, and people moved in and started farming. A new approach was needed. From 1976, a project funded by the UN Development Programme tackled an area of 58,000 hectares on the Subri River Forest Reserve. Instead of planting exotics, the project singled out wild saplings of valuable species, and cleared the area around them. The felled wood was not wasted. A mobile sawmill cut up timber for sale, logs and branches were converted to charcoal, and the remaining twigs and brushwood left as mulch. Local farmers were allowed to grow cassava and bananas among the young seedlings. The project conserved the forest and its wildlife, created local employment, and changed the relationship between farmers and foresters from conflict to co-operation. The initial costs were high, but even the short-term benefits from timber, charcoal and food production brought in a profit of $2,150 per hectare. On top of that, the tree growth rate was 45 per cent higher than under the old system.

The Subri cost of $6,000 per hectare is too expensive for most of Africa's forests: low cost and high rates of return must be a prime goal. The Guesselbodi experiment has shown that initial investments of $400–$500 per hectare can give a high return

(p. 146) in the case of severely degraded forest. For non-degraded forests, little more is needed than protection by firebreaks and guards, which can cost as little as $20 per hectare.

The science of natural forest management in Africa is in its infancy. Research is needed into the stock of species, how they are used by local people, and how much they produce each year. Potential markets for forest products have to be assessed. Methods have to be developed to improve productivity – what are the best and cheapest methods of protection, what type of coppicing or cutting is best, what rotation systems are needed.

Such research can take many years – but in many semi-arid areas immediate action is needed. The first step must be to alter the relationship between farmers and their local forest; to shift ownership, or at least the right to use, manage and protect the forest, from the state back to the villagers to whom the forests originally belonged. Where a suitable village organization does not exist, it may be necessary to create a forest users' association. There will still be a temptation for individuals to overexploit – but at least the community will have an incentive to try to stop them. Forestry staff should be retrained as forest extension workers. They will spread knowledge of proven techniques, but at first they will organize villagers to survey forest species, products and uses, and test out directly a few basic protection and improvement measures in collaboration with local people.

Capturing and developing this potential would obviously be much cheaper than plantations, more ecologically and culturally sound. In view of the vast areas and resources involved, it could also have a much greater impact on preserving the environment, providing fuelwood, food and other products, and creating jobs.

The second hopeful avenue lies in abandoning the blinkered conventional idea that forestry can only take place in forests. The central purpose of forestry in Africa should be to encourage the planting and sustainable use of trees everywhere. In some cases the trees will be in separate blocks – in natural forests, on watersheds and steep slopes, in urban fuelwood plantations. In others they will be widely scattered, in compounds, schools, and government offices, along rivers, roads and farm boundaries, and increasingly *within* fields, where trees' capacity to protect and enrich soil can have most impact.

BRAKING THE WIND

One of the most successful projects of this kind in Africa is in the Majjia Valley, in the department of Bouza, Niger. The valley's windbreaks are an arresting sight. Row after row of flourishing neem trees, waving their branches gracefully in the steady breeze, march like a well-drilled infantry regiment through the yellow fields of millet stalks. Starting in 1974 at the northern end of the valley, they have continued their unstoppable advance: 330 kilometres of windbreak in all, with 80 kilometres added in 1985 alone. By any standards, the project is a success: in the context of African forestry performance, a spectacular success.

The valley itself was heavily wooded at the beginning of the nineteenth century, when it was settled by Hausa farmers fleeing from the Tuareg who dominated the surrounding plateaux. The soil is fertile. In the wet season, sorghum grows on the lowest parts of the valley bottom, providing a short second crop after the first is cut. On higher land, millet is intercropped with cowpea. Floods cover parts of the valley in years of good rain, soaking the ground and bringing a coating of fine alluvium: crops planted when the waters have receded provide handsome harvests. In the dry season the water table remains high, and farmers dig shallow wells to irrigate potatoes, cassava, onions, tomatoes and tobacco.

The valley's fertility supports a population of 33,000, densely settled at 75 people for every square kilometre. But the land, like most of the farmland in Niger's agricultural zone, lies under threat. It is surrounded by slopes and plateaux of rusty-hued lateritic rocks and stones, stripped of almost all vegetation. Older villagers remember times when the slopes were covered in thickets, inhabited by antelopes, hyenas, monkeys, even lions. Persistent cutting, and the droughts of 1968–73, took their toll, so that by the mid-1970s the hills were almost completely bare, a dark red, barren, Mars-like landscape. The Majjia river now flows faster, carrying a heavier load of silt and sand. The bed has been scoured a couple of metres lower, so that the river no longer floods the plain in the valley's upper reaches.

From November to February the Harmattan wind blows hot from the north-east, laden with dust and sand that veil the landscape in a grey-tawny haze. The valley funnels the wind, which reaches speeds of up to 60 kilometres an hour in January. In the dry season, vegetation cover is sparse, and the topsoil blows away at the rate of 20 tonnes per hectare each year. In the rainy season the wind direction reverses and the problems change. The wind now covers emerging seedlings with sand, so farmers have to plant several times. It parches the soil, wilts the growing crops, and blows them down when they have headed.

The Majjia valley project began in 1974, when local villagers asked the government forestry officer in Bouza, Daouda Adamou, if he could do anything to help them against wind erosion. A few lines of windbreak had been planted a decade earlier, far down the valley: they provided the inspiration. Adamou and an American Peace Corps volunteer, Don Atkinson-Adams, approached the US relief and development agency CARE for funds. Work began in the dry season of 1974, with the setting up of the first nursery. Planting started at the beginning of the rains in the following year.

The species chosen for the windbreaks was neem, a deep-rooted evergreen from Asia. Neem grows well in semi-arid areas, and can survive, once established, on rainfall as low as 130mm per year. The wood provides a high-quality fuel and a tough, decay-resistant timber. Neem oil from the seeds can be used as lamp fuel, and the residual cake is an excellent fertilizer. The seeds and leaves yield azadirachtin, a very effective insect repellant. The neem was planted in double rows, with trees spaced four metres apart and 100 metres between one windbreak line and the next.

The Niger forestry service is a centralized, authoritarian body with a paramilitary style. Since all trees, in law, belong to the Niger government, foresters have powers to control felling and even pruning of trees on farmers' land: forest agents patrol on motorbike or horseback, and carry guns. Daouda Adamou pioneered a new approach to forestry in Niger. Villagers were viewed not as potential criminals to be kept in check, but as potential partners and principal beneficiaries of forestry work. Adamou prepared the ground for the project carefully, spending long nights in the villages educating people about the benefits of trees. But popular participation was, and still is, modest. The villagers dig the holes during the dry season, ready for the trees to go in. Foresters and Peace Corps volunteers choose the location of the windbreaks and mark the lines and tree positions on the ground. Paid nursery staff raise the seedlings.

Farmers are responsible for planting and looking after the trees in the rainy season. They weed them and protect them against livestock. In the dry season, when farmers are not often in their fields, the job is taken on by paid guards. Any animals they catch are impounded, and owners have to pay steep fines to release them – $25 for every camel or cow, $12.50 for donkeys, goats or sheep. Protected areas are marked by stakes with coloured rags: animals must stay out for three years, though farmers are allowed to collect forage and carry it to their beasts. Livestock owners are the main losers from the project. Many sold their animals, or moved them out of the valley, rather than pay the fines.

The sum total of benefits far outweighs the costs. The windbreaks take up land – perhaps 12–15 metres in every 100 metres, including the deep shade on each side of the row. But even after allowing for this loss, two separate studies have found that they raise crop yields by 18–23 per cent, compared with similar land unprotected by windbreaks. The windbreaks reduce soil erosion by lowering wind speeds below the threshold needed to lift and transport soil particles. They reduce the smothering of early seedlings by sand. They cut down evaporation and increase the level of moisture in the soil. Plants wilt less, and close their stomata less so there is more photosynthesis. There is less toppling of mature crops.

The villagers are well aware of the benefits. In a recent survey, as many as 72 per cent of farmers said the windbreaks resulted in increased food production – 54 per cent said they led to a significant increase. Abdullah Sami, a peasant from Garadoumé village, told me they would like more windbreaks: 'I can show you now the fields where we need them. The wind is blowing more soil away every year.' Farmers whose land is outside the current windbreak area express almost universal envy of those within. When CARE conducted a sociological survey of the valley, the first question people in villages not yet reached by the windbreaks asked the interviewers was, 'When will we be included?' The windbreaks have spread out from two centres, in the north and the centre of the valley. Tambalanga, a village stranded in the middle, began planting its own lines.

The additional benefits of the wood have only just begun to be reaped. In 1984 200 of the earliest trees to be planted were pollarded, and the wood was given to the village development council to distribute. It seemed a democractic, participatory method of distribution – but in practice local chiefs gave the wood to their family and clients. In 1985 a further 550 trees were cut and a new form of distribution was tried: one third was given to the owners of the fields where the trees stood, one third to the wood cutters, and one third to the village council. But village women descended on the cutting areas like a swarm of locusts, pilfering whatever they could get away with. During the night the cut wood was left without a guard, and every last stick of it disappeared.

CARE is making a very careful study of the Majjia valley project so the lessons can be learned. The first wood cuts were experimental. Varying methods were tried: coppicing (cutting at the base); pruning off all branches that overhang fields; and various patterns of pollarding – cutting at about 2.5 metres from the ground; both rows, one row only, or every fourth tree. Regrowth is prodigious – neem pollarded in June 1985 had, by November, put out shoots two metres long and five centimetres thick in a year with only 350mm of rainfall.

The wood is sorely needed. Firewood is in increasingly short supply in most parts of the valley. It is not uncommon for women to spend half the day scouring the hillsides for kindling, and half of all families have to buy firewood at least part of the time. Only the villages of Ayaouane and Keleme report ample supplies. Here the thorny leguminous tree *Prosopis juliflora*, planted as live fences and to protect the river banks, has run wild in rampant, verdant thickets. It grows so vigorously that villagers can hardly cut it fast enough, and provides rich supplies of fuel, poles, and fodder from the pods.

The villagers are impatient for ample, regular cuts of the windbreaks. Alhaji Adamou Addo had to spend $40 on firewood in 1984. In 1985 he got a one-third share of six trees that were pollarded in his fields; they provided enough poles to build an extra room for his compound, plus two months' fuel. I was curious to find out if he valued the trees enough to care for them himself, and asked him what he would do with them if the forestry department let him do as he liked. He said he would pollard them, all of them, and take the benefits – and they would grow again, rapidly. Last year's cut showed that, clearly. Would he go on cutting them if he saw the trees could not sustain it? No, he would only take as much as they could produce easily.

The project, together with the object lessons of two droughts and a growing fuel crisis, has changed local attitudes to trees. Along with two thirds of the valley's farmers, Alhaji Adamou fosters any useful wild trees that seed themselves in his fields in between the windbreaks. A number of farmers have started private woodlots. Adamou Gouja of Tama village owns only a quarter of a hectare of land. In 1984 he decided it would be much more profitable to grow wood for sale rather than crops, so he planted his plot with neem and eucalyptus seedlings. By November 1985 they were three metres tall. Around 60 farmers in the valley have started their own private tree nurseries and sell seedlings to their neighbours.

Contrary to what has sometimes been claimed, the Majjia valley project did not succeed because of a high level of popular participation. True, participation was much stronger than usual in Niger forestry, but the Majjia project was still planned by forestry officials and a foreign voluntary agency. It was executed largely by paid nursery men and guards (though these were recruited from local people). Windbreak lines were laid out with a cavalier disregard for – indeed in total ignorance of – their effects on the distribution of land. There was no compensation for farmers who might have lost a considerable area. The method of distributing cut wood was decided by officials, not locals, and has still not been fully resolved. When Majjia farmers are asked who owns the trees, they say simply, 'The forestry department.' Local attitudes to the project are summed up aptly by farmer Abdullah Sami: 'It was an initiative of the state, the people had no choice but to accept.'

And yet it did succeed. Though there was no formal consultation, the villagers had asked for help. Daouda Adamou's evening talks with villagers clearly gave him an insight into their circumstances, and the project did offer a solution to wind erosion, which was perceived as a serious problem by local villagers. The technology was appropriate and simple, the species well adapted to local conditions. The package delivered the protection that had been promised. The costs for the majority of farmers were slight. The benefit in increased crop production was rapid, and perceived by the farmers.

The project was well worth while to the local population. Because of the long wait, the benefits in wood alone would probably not have assured success, but they came as a free bonus, on top of the benefits for food production. Local people reaped very real advantages from the project, and they were aware of the advantages. This was enough to ensure their co-operation.

The project could be improved. Local people would be perfectly capable of running nurseries, either privately or communally, with a little training and material input. Equally, the farmers could easily plan the windbreak lines themselves: the task of laying out a straight line perpendicular to the wind and roughly parallel to another straight line 100 metres away should not require a forester. If farmers were guaranteed the full benefits of trees on their land, they would probably organize themselves to guard or protect the seedlings, as they do with valuable volunteer trees that sprout up. Lines of trees along the borders of plots can also act as windbreaks. They can be planted by the owners and do not require communal arrangements to install or manage.

Within the past two years, CARE has begun to spread the windbreaks to other areas, starting with the districts of Maradi and Birni N'konni. At Birni N'konni they will protect a 2,500-hectare irrigated area run by a farmers' co-operative. A number of the problems encountered in the Majjia will be avoided here. The co-operative will be responsible for managing, maintaining, harvesting and selling the wood from the windbreaks. Windbreaks of eucalyptus, *Leucaena* and *Prosopis juliflora* will be planted along canals and paths and round the perimeter, so no farmer will lose land to them.

A NATIONWIDE CRUSADE: TREE-PLANTING IN KENYA

In Niger, tree-planting is still largely restricted to donor-financed plantations, village woodlots, windbreaks and living fences. They are thinly scattered efforts, falling far short of the urgent demands of the situation. This is typical of most of Africa. Only a few countries have made breakthroughs in forestry on a large scale. In Burundi, Rwanda and Malawi, planting on individual farms has reached a level where some experts believe the deforestation problem may be well on the way to being solved.

In many parts of Kenya, the scale of tree-planting is prodigious. A survey carried out in 1983 by the Kenya Woodfuel Development Project, in Kakamega district, found that four out of five households had planted trees on their land in the previous twelve months. Some 64 per cent had their own woodlots. More than half of those who planted collected wild seeds or seedlings from the bush. Two out of five had their own tree nurseries.

It is not unusual to encounter Kenyan farmers who have an almost religious fervour for tree-planting. People like Stephen Wanje, who back in 1968 bought 4.5 acres of badly eroded land in Fudumi in western Kenya. He paid 20 shillings for it (about $1.30), but it was as hilly as a rollercoaster, and neighbours wondered why he was willing to pay so much. 'I began by building terraces, then I started planting trees. People laughed at me. "Why start planting trees?" they said.' The steepest slopes on Wanje's land are now dense plantations of native trees, and his family is more than self-sufficient in fuel and timber. 'Today,' he says with satisfaction, 'the same people who laughed at me come to me for poles when they want to build a latrine or a new roof.'

Wanje runs a clinic in nearby Maragoli. The farm is a passionate hobby, dedicated to the principles of intensive organic farming and recycling. Beehives shelter under the trees. In a hollow are four fish ponds. From the excavated clay, he makes bricks and tiles, fired with his own wood. On the small area of flatland he grows napier grass, a perennial fodder plant, to feed a herd of ten cross-bred cows. The animals' dung feeds a biogas plant, from which Wanje heats and lights his house; the residue provides a high-quality fertilizer. Wanje's farm is a regular stop for trainees from the farmers' training centre at Bukura. 'The sun is free, the rain is free, the earth is free,' he tells his visitors. 'With these, you can do anything. Whatever you see here, you yourselves can do. The ladder of progress for a farmer has no limit.'

Wanje is an educated, dedicated man with enough funds to invest in experimentation. But many ordinary farmers in Kenya had the foresight to value trees long before it became fashionable – people like Boaz Mukati, a white-haired 62-year-old who farms in a shirt that is torn almost to ribbons. Mukati owns a one-hectare plot of land near Bukura, on a 60° slope so steep it is hard to walk along. In his youth he grew food crops here, but in 1956 he was arrested by the colonial District Commissioner's officers for cultivating an excessive slope. He spent fourteen days in jail, and had to borrow to pay the hefty fine before he was released. He decided to plant eucalyptus on the land. The rainfall here is high, and trees put on up to 10 metres a year with little trouble. Eucalyptus can be coppiced every two years. Some of the original trees have been cut fourteen times, and are still sprouting new stems. Mukati sells an average of 1,500 trees every year, mainly as poles for house-building. At 8–10 shillings each, they fetch a handsome income of over 1,000 shillings a month – more than the wage of most Nairobi factory workers.

Mukati and Wanje were pioneers, but they are no longer exceptional in Kenya. There is a long-standing tradition of agroforestry. Farmers in Kenya allow certain tree species to flourish in their fields: *Sesbania sesban*, a small vigorous leguminous tree, useful for firewood; mangoes, which sprout up from household refuse and compost; and decorative trees like the yellow-flowered *Markhamia* and the glorious red tulip-blossomed Nandi flame. Most plots are fenced not with dead stakes or stalks, but with hedges of cypress and *Dracaena*, sisal and *Euphorbia*.

Deliberate, planned tree-planting is a more recent development. Only in the past five or ten years has it become a mass groundswell. It is now quite hard to come across farmers who are not keen to plant as many tree seeds or seedlings as they can lay their hands on. The supply of seedlings from private community and government nurseries is growing exponentially, but all agencies in the field report a demand that still far exceeds the supply. People dig up self-planted seedlings from the roadside. Farmers who plant valuable species sometimes have their saplings stolen. 'When I unload a jeep full of seedlings,' the foreman of one agroforestry centre told me, 'people flock around in crowds, like when there is an accident.'

Tree-planting, along with soil conservation, has reached the proportions of a national crusade in Kenya. The beginnings, in the early 1970s, were modest. In 1971 the Forestry Department launched its Rural Afforestation Extension Scheme, with an eventual aim (still not achieved in 1985) of one extension worker in each district, providing technical advice to community and local government nurseries. In 1975 the soil conservation programme was set up (p. 118); it now has 29 tree nurseries, and recommends planting of trees along terraces. In 1979 the Ministry of Energy was created. Its spending on woodfuel rose twenty-fold between 1982 and 1986, rising from a mere sideline to the largest item in the budget. In 1980 Kenya's President, Daniel Arap Moi, issued a directive that every district commissioner and every chief should have a nursery (in Kenya chiefs are local government officers). From 1983 every secondary school was required to have a tree nursery and a woodlot or agroforestry plot. In 1985 the Presidential Tree Fund was set up to encourage planting by the youth section of KANU, Kenya's sole political party, and all party branch chairmen were required to establish tree nurseries. By 1983 there were as many as 1,300 government tree nurseries with a massive stock of 83 million seedlings. In that year, no less than 3 million trees were planted during the National Soil and Water Conservation Week alone.

Kenya's flourishing voluntary sector has also played a strong part in pushing tree-planting. As many as 60 organizations are in the field. One of the most prominent is the Green Belt Movement set up by the National Council of Women of Kenya in 1977. The movement, headed by former anatomy professor, Mrs Wangari Maathai, helps communities to set up 'green belts' of at least 1,000 trees on open spaces, in school grounds, along roads. The Green Belt Movement was one of the first in the field, and has publicized and popularized tree-planting among women and among political leaders. There are now more than 1,000 green belts, and as many as 20,000 'mini-green belts' on farmers' fields, along with 65 community tree nurseries run by women's groups.

A number of obstacles still remain in Kenya. There is a great deal of duplication among the many voluntary and government agencies involved in forestry, with a number of different approaches, balances and ranges of species. That may be no bad thing at this stage. A few more years' experience will show more clearly which approaches are most effective. Eventually the government's own forestry activities – currently carried out by no fewer than seven separate ministries and agencies – will have to be knit into a structure that reaches every Kenyan farmer with supplies and advice, backed up with regional research and technical services. All the elements are already there: they simply need co-ordinating.

Another major problem is the position of women. Many husbands migrate to cities and large estates for work, so women make up at least one half of Kenya's working farmers. Yet they do not have title to their husband's land, and in many cases must ask permission to plant trees. In some areas, indeed, it is taboo for women to plant trees. The fact that fuelwood-gathering is women's work means that men often do not perceive the fuelwood shortage, and may plant trees for sale as poles or timber, even though their wives spend hours searching for firewood. Only when men have to shell out cash for firewood – or see an opportunity for selling firewood to others – would they plant trees primarily for fuel.

Nevertheless, the whole Kenyan push for reafforestation must be seen as an on-going success. The reasons for that success are replicable. Part of the secret lies in the laws of land and tree tenure. Most Kenyan smallholders enjoy full ownership and control over their land and their trees. They can plant what they like, and cannily shift production from one crop to another in response to changes in costs, prices and taxes. In this situation wood shortages can be self-limiting: a rise in woodfuel prices leads almost automatically to an increase in tree-planting. Of course, not all countries may wish to go for full private ownership of land. The state or the community may retain formal title, but what really matters is that farmers should have security of tenure, full control over planting and management of the trees on the land they use, and full enjoyment of their products.

Another key factor in Kenya's success is political commitment, starting at presidential level and increasingly embracing leadership at all levels. That commitment is not just verbal: it is backed up by resources and a real shift in spending priorities. And the message of tree-planting is broadcast through every available channel, in speeches, in newspapers, on radio and TV, through schools, extension workers, local government chiefs, the political party, and voluntary groups. As George Mburathi, chief executive of the Permanent Presidential Commission on Soil and Water Conservation, puts it, 'There is scarcely a single person in Kenya today who has not heard of the importance of tree-planting.'

One of the most encouraging signs is that even those farmers who were part of the deforestation problem a couple of years back are now beginning to search for solutions. I met Pius Mbithi sitting on a pile of eight sacks of charcoal, stacked by the roadside in the semi-arid region of Machakos, waiting for a passing purchaser. Charcoal-making is virtually his only source of cash income. To pay for school uniform and school fees for just one child, he must sell thirteen bags of charcoal – equivalent to a couple of sizeable trees. Mbithi has a farm of eight acres, growing meagre crops of maize and pigeonpeas. He has progressively felled all but six trees on his fields for charcoal-making – many of them in the drought year of 1983–4 when he had to buy in extra food. With his cash resource visibly dwindling before his eyes, he has been forced to think of tree-planting. In 1984 he planted 45 seedlings. Only one survived. The rest were eaten by goats and baboons, or killed by termites attacking the roots.

Kenya is doing well, but there is no room for complacency. The task ahead is to create the nationwide supply and extension system that can provide people like Pius Mbithi with seeds or seedlings, and the guidance they need on rearing them.

11 Turning farms into forests: the potential of agroforestry

In the past, forestry has been considered separate from agriculture and livestock, concerned solely with managing plantations and forests producing timber and fuelwood. Relations with peasants were hostile. Foresters viewed farmers and herders as vandals, destroyers of forests, to be kept out at all costs. Peasants saw foresters as policemen who excluded them from land that was traditionally theirs to control and use. Under such conditions the forests did not flourish, and farmers came to view tree-planting as an alien activity carried on by unpopular professionals.

Yet forestry has a crucial role in the future of farming and pastoralism in Africa. It can *only* play that role if it is fully integrated with crop and livestock production and trees are planted on a vast scale, on farmland and pasture. Then their benefits are not restricted to plantation and woodlot areas, but spread over all the land that needs their protection.

Trees on and around farms meet much wider needs than woodlots and plantations. They serve as windbreaks and shelter-belts. They mark boundaries and strengthen terraces. They provide shade, ornament and privacy for homesteads. They supply not only fuel, timber, stakes and poles, but also cash crops, fodder, fruit, nuts, oilseeds, leaves and pods for sauces, tannin, dyes, gums, resins, fibres and medicines. Fodder and food trees can help to balance diets, and to fill the 'hungry gap' for animals and people at the end of the dry season and the beginning of the rains, when other food sources are scarce. They also act as drought insurance: with their deeper roots, they are more likely to yield something in dry years when conventional food crops fail. Perhaps their most crucial role is in recycling nutrients, in a climate where heavy rains quickly leach them down below the reach of crop roots, and in maintaining the level of organic matter in the soil, in an environment where high temperatures break down organic matter too quickly.

Agroforestry – tree-planting on and around farms – offers by far the speediest road to reforesting Africa. Professional foresters are few, but agroforestry could convert all of Africa's 35 million smallholders into potential foresters. Fostering planting by farmers on their own land is much cheaper and simpler than finding sites for village woodlots and making arrangements for collective sharing of workloads and wood production. The trees are more likely to survive and thrive, because farmers will invest more time and effort in tending and guarding the seedlings. Financial returns on investment are much more attractive. Costs are lower, and total benefits, including all tree products and environmental protection, are higher. A 1983 study of World Bank forestry projects found that plantations gave average returns of 10–15 per cent on investments of $800–$1,500 per hectare. Farm forestry projects cost a quarter to one third as much, and earned returns of 25–30 per cent.

Many African farmers and herders already practise one form or another of agroforestry, though it is not always recognized as such. The most widespread practice is the use of trees growing on fallow land. Most families have shade or fruit trees in their compounds. Many farmers foster useful seedlings that sprout spontaneously in their fields. Many grow tree and shrub crops, from oil palm and cocoa, to coffee and tea, bananas or plantains.

But outside the compound or the tree-crop plot, most African peasants have relied on natural processes of self-planting to provide them with trees. These processes are now under severe threat. As fallow periods are cut down, trees may no longer have time to reach maturity and start producing seed. Wild seedlings face risks from overgrazing.

The task ahead then is to convert African farmers and herders from *passive* to *active* agroforestry; from users of self-planted trees, to tree-farmers. It is the equivalent of the transition from gathering wild cereals, to planting them: an agroforestry revolution, to follow up the agricultural revolution.

A number of approaches to agroforestry have already proved themselves in practice. In drier areas the leguminous tree *Acacia albida* (p. 75) is an ideal companion for crops and livestock and will tolerate rainfall as low as 300mm a year where deep groundwater is available. In Sudan and Senegal, the legume *Acacia senegal* is used in an organized 20-year rotation of crops and grazing. In the first five years, millet is grown among the young trees. Over the following fifteen years, controlled cutting of fodder and grazing is allowed. At the same time the trees are tapped for gum arabic, a valuable product used in medicines, sweets, soft drinks and printing. In the final years mature trees are felled and used for fuelwood and charcoal. The taproot can be made into tool handles and weavers' shuttles, while the long surface roots provide strong fibre. Seedlings and millet are planted on the cleared ground, and the whole cycle begins again.

FERTILIZER FROM TREES: ALLEY CROPPING

The most promising approach to agroforestry in Africa is a revolutionary technique known as alley cropping. It recognizes the key role of trees and shrubs in maintaining soil fertility, but boldly shifts the trees from the fallow area to the crop field. In alley cropping, food crops are grown between pruned hedgerows of fast-growing, nitrogen-fixing trees. The leaves are used as mulch, and act as free fertilizer, improving soil structure and boosting crop yields by upwards of 35 per cent.

Alley cropping is a superior successor to fallowing and shifting cultivation in a time of booming populations. It allows the same piece of ground to be cultivated continuously without fertilizer, and improves rather than degrades the soil. Alley cropping has been developed over the past decade at the International Institute of Tropical Agriculture in Ibadan, Nigeria, by an Indonesian soil scientist, B. T. Kang, and colleagues. 'We learned the hard way here,' Kang remembers. 'When I arrived at IITA, I thought we knew the answers to Africa's problems. We made contour bunds, but we still got tremendous erosion. We used chemical fertilizer, but the soil grew more acid. Then we started to look a lot more carefully at what African farmers themselves were doing. We found they were using trees and shrubs, in the fallow period, to restore soil fertility. But the problem with the fallow system is that it stays at a very low level of productivity. Yields fall off steeply if the plot is cultivated beyond the first year. So we asked, "Can we improve the traditional system? Can we organize it?" '

Kang was familiar with eastern Nigeria, where farmers, forced into intensive techniques by population density, had integrated trees into their arable farming (p. 76). Kang had the idea of systematizing and simplifying the practice so it was easier to research and teach to farmers. The result was alley cropping. The first field trials at Ibadan started in 1976.

The trees used in alley cropping must be carefully selected. They have to have deep roots, so they do not compete with food crops for water and nutrients. They should be fast-growing, able to sprout again easily after coppicing or pollarding. They should ideally be multi-purpose trees, providing fodder or mulch from their leaves, fuelwood and stakes from their stems. They should preferably be leguminous, able to fix their own nitrogen, so they provide protein-rich leaves for livestock, and nitrogen-rich organic matter for the soil. A number of trees fit the bill well. The leader is *Leucaena leucocephala*, a native of southern Mexico. It grows fast, even on marginal soils, and provides high-quality fodder, mulch, and firewood. Other promising species are *Cassia siamea*, a fast-growing firewood tree, *Gliricidia sepium*, *Calliandra calothyrsus*, and *Sesbania sesban*, all legumes.

The management is straightforward, and can be mastered by farmers with very little instruction. The tree seeds or seedlings are planted as close together as four to ten per metre, in rows 4–8 metres apart. They can be planted at the same time as the first food crop, and weeded and protected along with the crop. By harvest time, the trees are big enough, in humid areas, to be less vulnerable to browsing goats. In the dry season, they can put on as much as another four metres on residual soil moisture. The leaves can be cut for fodder in the lean months.

Just before the next rainy season, the trees are pruned to anything between 25cm and 2 metres from the ground, so they don't shade the growing crops. The larger stems can be used as stakes or firewood. Twigs and foliage are spread on the soil as a mulch, or dug into the soil. The trees are pruned again every 5–6 weeks during the cropping season, up to five times a year.

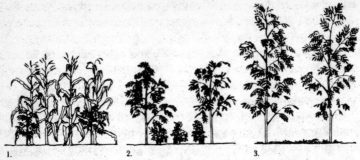

Alley cropping with leguminous trees like *Leucaena*. In the first year the
seedlings are planted with the crops, (1, 2) then allowed to shoot up in
the dry season (3)

In subsequent years, the trees are pruned for wood, fodder and mulch
while crops are growing, (4, 5) and allowed their head again in the dry
season (6)

They can produce up to 6 tonnes of stakes per hectare each year,
and 15–20 tonnes of leaves and twigs for mulch or fodder.

Although the tree stems may take up 5–10 per cent of the
land, there is no loss of food output; indeed, crop yields increase
handsomely because the leaves of leguminous trees are valuable
fertilizer. *Leucaena* leaves can provide as much as 166 kilos of
nitrogen, 150 kilos of potash, and 15 kilos of phosphorus per
hectare. When incorporated in the soil, every tonne of leaves

produces the same increase in maize yields as 10 kilos or more of chemical nitrogenous fertilizer. *Leucaena* leaves greatly increase the efficiency of chemical fertilizers. When the prunings were combined with 40 kilos of nitrogenous fertilizer per hectare, they gave a greater increase in yield than a massive dose of 160 kilos of chemical fertilizer alone.

On-station yields from alley cropping have been impressive. At Ibadan, rice grown between rows of *Sesbania rostrata*, and mulched with 4 tonnes of leaves per hectare, yielded 36–55 per cent more than a control plot. 120 kilos of chemical fertilizer would be needed to achieve a similar effect. The yield advantages of alley cropping build up with time, as the organic content of the soil increases. In a long-term trial on a sandy soil in Nigeria, unfertilized maize mulched with *Leucaena* prunings yielded 83 per cent more than without: 1.9 tonnes per hectare, as against 1.04 tonnes. In the following three years, the output of the untreated plot fell to an average of only 0.45 tonnes of maize per hectare. The alley-cropped plot went on yielding at around 2 tonnes per hectare, except for one drought year – an average of three times more than the untreated plot, and more than double the average yield in Nigeria.

Like bush fallow, alley cropping provides fodder, fuelwood, stakes and poles. But after fallowing, the nitrogen and other nutrients built up in the soil are available for crops only in the first year or two: with alley cropping, they are available permanently. Bush fallow reduces soil erosion in the fallow area; alley cropping reduces it in the cropped area normally exposed to wind and water erosion. Indeed, densely planted hedgerows, along the contours of moderate slopes, collect soil on their upper sides, and terraces build themselves within a few years.

Many panaceas for Africa's problems have not survived the crucial move from the research station to the farmers' fields. The signs after four or five years of on-farm trials of alley cropping are very encouraging. Farmers have achieved increases in maize yields averaging 39 per cent.

The largest farmer trials so far have been carried out by the International Livestock Centre for Africa. In 1981 ILCA's humid zone programme moved into the IITA campus at Ibadan, and began testing *Leucaena* and another legume tree, *Gliricidia*, as fodder for goats and sheep. The leaves of both species are very high in protein. Animals that were given only 800 grammes of leaves a day as a food supplement spurted ahead of the rest. Ewes gave birth more frequently, lambs put weight on faster, and more survived to weaning. The overall output, in kilos of weaned lamb per ewe per year, rose by 55 per cent.

In 1981 the ILCA team set up demonstration plots in two Nigerian villages near Oyo, Owu-Ile and Iwo-Ate, in the border zone where forest gives way to savannah. In 1984 they asked for volunteers to take up alley cropping; no less than 68 farmers, a quarter of all local farmers, came forward. There were teething troubles. In one village, a credit programme was promised, then withdrawn, and several disgruntled farmers dropped out. The time for planting the tree seedlings coincided with the palm fruit harvest, a busy time for women, so not many women participated. Local attitudes to trees caused occasional problems. Land is, by custom, communally owned, though individuals have rights of use. Whoever plants a tree acquires a stronger hold on their land. Some farmers were afraid that ILCA, because it had provided the seedlings, would come to own their land. One landlord was so incensed when his tenant planted trees that he hacked them all down. The advice provided was none too detailed, and the follow-up weak. Farmers were very much left to their own devices. Some of them started collecting fodder from the saplings too early. On one woman's plot I saw, the trees were thin and stunted, or dead, after 18 months. In other cases labourers hired to weed the plots unwittingly weeded out the seedlings.

In spite of all these hiccups, when the alley farms were assessed in August 1984, four months after planting, more than half were judged to be good or excellent, and less than one in five poor. By November 1985, when I visited the area, some of the plots looked better than those on the research station. The best belonged to Akanni Bakare, a resourceful, innovative character who also uses an improved cassava variety and has invested in a motorized mini-mill for cassava, which he drives from house to house on his motorbike. His *Leucaena* and *Gliricidia* were about 4 metres high. He was just making ridges to plant yams, pruning the branches and leaves from the hedges and piling them in the furrows.

The soil was friable, rich in organic debris. Bakare used to cultivate one plot for six years, then leave it fallow for four. 'With the trees, I don't think I'll have to leave this plot fallow,' he told me. 'It seems as if the soil is much better. If it were not for the trees, the soil would be hard now. It would be much harder work to make the ridges.' Bakare used the poles from the trees as stakes for his yam vines to grow up. Staking increases yam yields by 100–200 per cent because the leaves get more light, but many Nigerian farmers cannot find suitable poles and no longer stake their yams. Bakare takes foliage home every day for his goats, and they gather expectantly when they hear his footsteps. The extra fodder has allowed him to increase his flock from three to five, and he says the kids look stronger. The example of farmers like Bakare has convinced many more villagers to join in. Alley farmers have given and sold tree seeds to neighbours. In the 1986 season at least 100 farmers will be alley farming.

Alley cropping is well suited to Africa's social and economic setting. It costs little to set up, especially if easily available species are grown from seed. No benefits are forgone – though a small amount of cropland is used for trees, crop yields actually increase from the first year in humid areas, and the second in sub-humid. The wood and fodder production is pure bonus. The system costs no cash to keep going, and reduces the need to spend cash on fertilizer. It is easy to learn, and easy to teach. It is flexible – farmers can vary the emphasis on fuel, fodder, or food to suit their own priorities and can go on farming crops in their traditional ways. Trial farmers, in typical African fashion, are already beginning to experiment and vary the basic system.

The only major question mark about alley cropping is whether it is too labour-intensive, given Africa's widespread labour shortage. When devising alley cropping packages to recommend to farmers, the labour involved should be minimized. The rows can be planted, weeded and guarded at the same time as the food crops. Surface mulching with the leaves involves far less labour than digging them in – but any digging in could be done at the same time as the land is prepared for the crop, or weeded. That still leaves the work of pruning the hedges, and clipping off foliage and twigs. However, this labour yields stakes and firewood, which would take far longer to gather in the bush or forest, plus fodder which increases livestock output and, as one canny chief put it, 'reduces the competition for food between goats and people'. In addition, alley cropping reduces the labour of weeding – weeds have a hard time getting going before the alleys are pruned, and mulch suppresses them. Akanni Bakare found it makes the soil easier to work, and eliminates the laborious work of clearing fallow land every year or two. Model costings have shown that even if labour requirements are 50 per cent up on traditional methods, alley cropping would raise profits per hectare by 60–100 per cent, and income per hour of labour by 7–33 per cent.

Alley cropping is ideally suited to humid and sub-humid zones. For those areas, it has now proved itself in station and farm trials, and is ready for wider dissemination through packages designed in co-operation with farmers. *As of early 1986, there was not a single large-scale or national programme in Africa using alley cropping, and it is high time a start was made.*

In theory, alley cropping should be adaptable for semi-arid zones by using different tree species, and by increasing the spacing between rows or within rows. *Leucaena* will grow with as little as 600mm per year of rainfall. At one drylands trial in India, alley cropping with *Leucaena* increased sorghum yields by 96 per cent. *Sesbania sesban*, a legume that grows widely in Africa, can tolerate as little as 350mm, and can withstand acid soils, waterlogging and flooding. *Acacia* and *Prosopis* species may also be suitable. Below a certain rainfall threshold, more widely spaced windbreaks may be the best approach. All these are open questions. The fact is that the basic research in alley cropping for semi-arid zones in Africa has hardly begun. That research should be a top priority.

A NATIONAL AGROFORESTRY NETWORK

Agroforestry as a science is still young. The research agenda to be covered is vast. Agroforestry involves the interaction of trees and crops, and often livestock too, in three dimensions above and below ground, with additional complications of culture, society and economy. The permutations are almost infinite. Inevitably it will be the farmers who develop the detailed combinations suitable for their specific environment, their needs and possibilities. It is a complex intercropping problem of the kind that the African peasant is expert at.

But there is an important role for formal research, in identifying broad approaches, selecting the most productive species, breeding improved varieties, testing out management methods. Despite its importance, agroforestry research is a very impoverished cousin of conventional agricultural research. It is grossly underfunded and understaffed. The International Council for Research in Agroforestry, based in Nairobi, has so far concentrated on developing research methods and information exchange, rather than on substantive research. The International Institute of Tropical Agriculture still has only one professional-level researcher working full-time on agroforestry: B. T. Kang, the originator of alley cropping, who is a soil scientist by training. The Sahelian centre of the International Crop Research Institute for the Semi-Arid Tropics did not appoint its first researcher in agroforestry until 1986.

Since 1982, Kenya has had what is probably the most developed agroforestry research network in Africa, based in six agroforestry centres run by the Ministry of Energy. The centres were set up as part of the USAID-funded Kenya Renewable Energy Development Project. The project, managed on contract by a US firm, Energy Development International, was launched in 1981, in the wake of the second wave of oil price rises and a drought that threatened Kenya's hydroelectric power dams. It was originally intended to reduce the country's dependence on imported fuels, but has focused on improving woodfuel supply, and reducing demand by way of fuel-efficient stoves (p. 210).

To stretch the modest five-year $4.8 million budget, Ethiopian project director Amare Getahun was determined that he would not duplicate any existing equipment or services. So he decided not to build separate facilities, but to locate the agroforestry centres at the Ministry of Agriculture's Farmer Training Centres, and the project offices inside the Ministry of Energy. This canny move ensured that agroforestry quite soon came to be viewed as a key area for both agriculture and energy. Each of the centres represents one of Kenya's main ecological zones. There is one on the coast, one in a semi-arid area, one in savannah, and three in different parts of the highlands where most Kenyans live.

The centres' main functions are research and training. They screen tree species and varieties to see which grow best in each zone, and try out different types of tree spacing and harvesting. Already the research is beginning to uncover much more productive approaches. Most conventional forestry is based on felling the whole tree, and replanting, but pollarding and coppicing methods appear to yield up to ten times more. Again, foresters usually recommend planting trees at least 3 metres apart, but Kenyan research has shown that gaps as small as 50cm in the row produce a very much higher yield of wood. The reason is that seedlings are forced to compete fiercely with one another. They put down deeper roots and send up straighter stems, with fewer low branches. The wood is easier to harvest, and more saleable as stakes or poles. Where timber or fruit is the aim, spacings of 2–4 metres are better. All the trees in the research plots are grown in association with crops, to find the best tree and crop combinations and the best ways of managing the two to provide maximum income. For optimum crop growth the best spacing between tree rows seems to be 4–8 metres, with the rows aligned east to west to minimize shading. The spacing of trees *within* the rows makes little difference to crop growth. Trees grown in closely planted rows, between alleys of crops, yield as well as, and in many cases two or three times more than, conventional forestry plantations of the same area. The food crop spurs tree seedlings to grow faster to escape the shade. And the trees' roots capture and use any leached fertilizer applied to the crop, which would otherwise be wasted.

Another function of the centres is training farmers, extension workers, officials and students at the agricultural training centres and colleges to which they are attached. Training is important not only in transmitting practical skills, but in spreading the unfamiliar concept of agroforestry among agriculturalists, foresters and administrators.

The third function is extension and outreach to spread agroforestry among local farmers. The centres supply seeds or seedlings free of charge to community groups, schools and model farmers. Other people pay 0.2–0.3 shillings per seedling – and 8 shillings (about $0.50) for a grafted fruit tree seedling. Many African forestry ventures concentrate on a small number of exotic species, but the agroforestry centres stock a very wide range of fruit, fodder, timber and firewood, ornamental and multi-purpose trees – 127 species in all. The choice of what to take is left to the farmer. Most farmers start by taking species they know well, with a few unfamiliar ones to try out. But gradually the balance shifts: citrus, eucalyptus and *Leucaena* are now the most popular among regular clients. The centres try to balance the supply of different species with the demand, so the farmers' preferences directly influence the type of trees on offer. To spread the idea of agroforestry, the centres set up demonstration plots with volunteer 'model farmers', innovative personalities who are likely to influence neighbours. Some of the centres work with as many as 1800 women's groups, farmers' clubs, and schools.

The model farmers and contact groups provide clear proof that agroforestry is practical and attractive to African smallholders. Caleb Wandle is a model farmer for the Bukura centre in Kakamega district. Wandle has to survive on only one hectare, and uses it very intensively. In one corner of his plot he has a small woodlot of eucalyptus, enough to supply his own fuel needs with some over to sell. Some he uses to fire bricks, made from clay dug from a trench in front of his house. He grows beans intercropped with maize, and gets up to 2 tonnes of maize from half a hectare in a good year. He uses no chemical fertilizer, but sows maize seeds in individual holes filled with rotted manure. He has dug his own terraces, and planted them with high-yielding napier grass, for fodder. His two milking cows, permanently tethered by the house, give 5 litres of milk a day and provide a useful cash income of $100 a month. In April 1985, the extension worker at the Bukura centre persuaded Wandle to plant *Grevillea* and *Sesbania* seedlings between his coffee bushes. Just six months later the trees were 4 metres high. They provide shade so the sun does not burn the coffee leaves. Wandle uses the ground layer under the coffee to grow beans and tomatoes. Some of Wandle's neighbours have already started to follow his example.

Further afield, in Mumias, the Wekhonye women's group has been an even keener client of the Bukura centre. Mumias is in the heart of Kenya's sugar-growing area. Back in the early 1970s the new sugar factory at Mumias approached local farmers, offering to plough, plant, fertilize and harvest the sugar cane and pay for it. All the farmer had to do was clear the land and weed the young cane. The lure of easy cash proved irresistible, and most farmers gave over nearly all their land to sugar. The real costs did not show till later. Sugar takes two years to mature at this altitude. In the first year farmers had very little maize of their own, and no cash income to buy grain. As a result, malnutrition was widespread.

Thatching grass for roofing used to grow on the boundaries between small fields, but the sugar company wanted large fields and thatching grass grew scarce. The Wekhonye women's group started up in 1975 in response to the problem. They formed a savings club, and as funds accumulated, individual members were given lump sums to buy tin roofs. Five years later they turned their attention to a third side-effect of sugar. All their trees had been cleared from the fields to remove shading and to make clear runs for the tractors. A desperate fuelwood shortage developed around Mumias, and farmers had to spend $6–12 a month on firewood.

The group approached the agroforestry extension worker, and in June 1984 started their own tree nursery on land loaned by the group's tall, bearded vice-chairman, Simon Wanga (many women's group members, throughout Kenya, are men). Members dug a borehole for water, made shades of bamboo and thatch, and raised 20,000 seedlings from seed. They gave 3,000 seedlings to group members, and sold the rest for a total of around $300, which was more than enough to cover their expenses.

Simon Wanga has gone over totally to agroforestry. He has *Sesbania sesban* growing around every field boundary, groves of *Calliandra* and *Croton* for fuelwood, and a quarter-acre alley cropping plot where sweet potatoes grow between rows of *Markhamia* and groundnuts beside *Grevillea*. When weeding, Wanga heaps the trash into a line underneath the trees: within a couple of years, this will convert his sloping patch into a series of stepped terraces.

A NEW PRIORITY

New species and new seeds spread readily in Africa; new ways of managing crops and trees spread much more slowly. Planting, raising and managing trees are new and unfamiliar skills in most parts of Africa. They will have to be actively promoted if they are to spread fast enough to halt the degradation of cropland.

Agroforestry must come to be seen as an integral part of the business of farming, not as a separate activity. The best channel, as with soil and water conservation, is the single multi-purpose agricultural extension officer at village level, working together with farmers' groups, rather than individual farmers.

Extension workers should begin by discovering what trees local people value, and how they manage and exploit them. Then they should identify the key problems that villagers perceive as serious, and find ways that agroforestry can help solve these problems. In some places fuelwood shortage may be the overriding concern: in others, wind erosion, or lack of stakes for yams, poles for roofing, or branches for fencing. Increased production of food and fodder will be a priority everywhere, and this will usually prove the most promising entry point for agroforestry. Growing fodder for goats was the chief reason why villagers around Oyo began alley cropping; the other benefits became clearer once they had started up. The best prospect of all will open up when multi-purpose trees are used – or a mix of species that provides fodder, fuel, poles, fruit or fibres.

As with soil conservation, agroforestry methods should be designed together with local farmers. They should be simple, easily taught and mastered, based where possible on species and systems that are already known. They should cut down the need for extra labour to a minimum, and maximize the role of self-help and local self-management.

So far, even in the best programmes, there has been an overemphasis on centralized nurseries growing seedlings in little plastic bags and supplying them to farmers. This approach relies too much on outside inputs, and expects farmers to trek long distances, to come home with at most 40 seedlings on their heads, many of which will not survive. The emphasis in future should be much more on teaching farmers and farmers' groups to collect their own seed and raise their own seedlings in the shade of the family compound, where waste water is usually available, or to plant seeds directly on-site along with the crops, at the beginning of the rainy season.

A programme based on fostering self-help in agroforestry, with the backing of a national extension network, is within the reach of every African country. Agroforestry is not only the most promising approach to reafforestation and the supply of fuel-wood, it is also, in yield-boosting forms like windbreaks and alley cropping, the most hopeful avenue for intensifying African agriculture over the next five to ten years, increasing food production *and* reducing exposure to drought with few or no outside or imported inputs.

Agroforestry is arguably the single most important discipline for the future of sustainable development in Africa. It should be given the priority and resources that it deserves both nationally and internationally.

12   Of sticks and stoves: solving the fuelwood crisis

In the humid parts of Africa wood is so plentiful it is blatantly wasted. I remember, in the rainforest of the Ivory Coast, coming across a smouldering wilderness of ashes, charred trunks and blackened stumps, all that was left of an area of forest that had been fired to ease the labour of clearing the land. The wood that had gone up in smoke could have kept a hundred families in firewood for a year.

In many of the drier areas the story is quite different. In the overcrowded Kunzwi area, near Harare in Zimbabwe, only a scatter of sizeable trees survive, most of those reduced by amputation to one or two limbs. Women have to make a round trip of 20 kilometres twice a week to gather wood from uncultivated land on a white commercial farm. In the Rift Valley of Kenya, below the humpbacked Ngong hills, Maasai women spend up to four hours a day scouring the grassy plains for fuel. In Alesedestu, at the foot of the Bale mountains in Ethiopia, villagers have a Hobson's choice: they can get free wood from a forest 30 kilometres away, a journey that takes a day and a half with a donkey cart, or buy their wood from a market 13 kilometres away. The price: $1.50 for a bundle that can be stretched out for a week if burned with cowdung.

Wood has many uses in Africa. Poles form the framework of thatch roots, and often, too, the skeleton onto which mud walls are plastered. Branches are used for fencing in compounds, or dense palisades around the kraals where cattle are kept at night. Straight six-foot stakes are used to support yams in Nigeria. For every use there is a corresponding crisis: stakes and poles often run out before small branches that can be used for firewood. In Ethiopia's recent villageization drive, farmers all over the country built themselves new houses in compact settlements – many villages in the central Highlands had to import the long, thick eucalyptus stems they needed at great expense from hundreds of kilometres away.

The most serious and most far-reaching of Africa's wood problems is the fuelwood crisis. A recent FAO survey found that as many as 16 of the 45 countries in sub-Saharan Africa faced fuelwood deficits on part or all of their territory, and could meet their needs only by cutting trees faster than they are growing, and reducing the stock (see map p. 354). Another 18 countries were suffering an acute scarcity, where they could not supply their needs even by overcutting, and had to go short.

Wood is the main energy source for nine out of every ten Africans. It accounts for more than 58 per cent of total energy consumption in the continent as a whole – four times its share in Asia, and eight times the share in Latin America. For rural people, wood supplies over 90 per cent of energy. Yet out of the 1980 regional population of 320 millions, some 55 millions faced an acute scarcity, and 146 millions a deficit.

The fuelwood crisis imposes a growing burden on African families. In urban areas the impact is financial: in cities as diverse as Ouagadougou, Bujumbura and Nairobi, poor families have to spend 20–30 per cent of their incomes on wood or charcoal. That inevitably means inadequately cooked meals and health hazards, or sacrifices in other basic needs such as food, clothing or housing. In a growing number of rural areas, fuelwood is becoming a marketed commodity that families – usually those with inadequate land – have to pay increasing prices for. But for most rural families fuelwood is still free, and the burden is the growing number of hours that have to be spent searching for sticks and branches. Throughout most of black Africa, collecting fuel is women's work, despite the fact that bundles may weigh up to 50 kilos. This task, along with water-fetching, can consume 400–500 calories a day, out of an intake that is already inadequate. It is a major factor in malnutrition, and a major contributor to the shortage of labour for farming.

In most parts of Africa fuelwood shortage is the result rather than the cause of deforestation. But once deforestation has passed a certain threshold, cutting for fuelwood can become a serious threat to the environment. Many larger towns in the drier regions are surrounded by rings of deforestation spreading out like ripples in a pond. The shadow of major cities reaches deep into their hinterland. Khartoum's charcoal supply lines extend up to 200 kilometres away, and the frontier drifts southwards by fifteen to twenty kilometres each year. Nairobi's web stretches out as far as 300 kilometres, into the semi-arid areas that can least afford to lose their more limited tree cover. The results are familiar: increased erosion and run-off, declining water in the soil, declining yields. Once fuelwood supplies disappear altogether – as they have done in parts of Niger and Ethiopia – crop residues and even dung are burned instead of being returned to the soil. Soil fertility and water-holding capacity decline further.

Where populations are growing fast, and tree stocks dwindling, a process of runaway deforestation and fuelwood shortage can set in. Predictions based on projecting current trends in wood production and consumption are alarming. In Kenya, for example, wood supply was sufficient, in 1980, to meet demand, but 30 per cent of the demand was being met by depleting the tree stock. As the stock dwindles, the sustainable yield declines, so more and more fuelwood has to be obtained by cutting into the tree stock. If these trends persist, demand will exceed supply by 87 per cent in 1995, and by 185 per cent by the end of the century. Using similar predictions, World Bank foresters Robert Fishwick and Dennis Anderson claim that by the year 2000 deforestation in the Sahel will accelerate from its present level of 3 per cent a year to 13 per cent. In the Sudan, less than a third of the 1980 stock of trees would be left by 2000 and annual consumption would be ten times as much as tree growth.

In most countries, these alarmist scenarios may well not materialize. As in Kenya, rising wood prices usually force people to economize in fuel use, or to switch to alternative fuels, and the amount of tree-planting will increase as more farmers plant trees for their own needs or for sale. But these free-market mechanisms cannot be counted on to work reliably in Africa. Forestry regulations and customary land tenure interfere radically, preventing people planting trees soon enough to avoid environmental damage. In Ethiopia, cruel feudalism followed by state ownership of all land and forests left the peasant with little real initiative: as a result, the northern part of the country is almost totally deforested and erosion has reached murderous rates. At some point, we can be certain that action will be taken in all countries, either by individuals or the state. But in many cases catastrophic damage may have to occur first. It is clear that in many countries the worst of the fuelwood crisis, and its most serious impact on the environment, still lies ahead.

The crisis is complex: there is no simple single course of action that can solve it. Like all shortages, it needs to be solved by a combination of increasing supply and reducing demand. Demand can be reduced by shifting to alternative sources of energy. In some places like Nigeria or the Ethiopian capital Addis Ababa, fuelwood has become more expensive than kerosene, and kerosene is increasingly used. In the longer run, biogas, wind and solar energy will be more widely used, but all three are waiting for rock-bottom cheap technologies, adapted to local circumstances, before they can spread in Africa. Biogas is particularly attractive in theory: those areas that currently burn their cowdung would no longer need to choose between energy and fertilizer uses – they could have both. In practice biogas will face tremendous obstacles in Africa due to the separation of cattle-rearing and agriculture. It has a brighter future in the highlands of East Africa, but only when much cheaper designs come into use (p. 238).

In the short and medium term, fuelwood is likely to remain the chief source of energy, so increasing the *supply* is paramount. This means, quite simply, planting more trees. Not just fuelwood trees, perhaps not even primarily fuelwood trees. Multi-purpose trees, providing timber, fodder, fruit, mulch, hedging or windbreaks, are likely to prove more attractive: all of them automatically increase the potential supply of fuelwood from prunings, waste or dead trees.

The most promising approach to reducing *demand* for fuelwood is to increase the efficiency with which it is used. Traditional wood fires in Africa are usually made of three stones. They are zero-cost, simplicity itself to build, and flexible enough to fit any size of pot. They give light, and provide heat in the areas that need it. The smoke kills insects, moulds and fungi, prolongs the useful life of thatched roofs, and preserves foods stored above it. But the traditional fire has serious drawbacks. Most African huts have no chimneys or windows. The choking smoke causes eye and lung complaints and probably cancers as well. The fire is also inefficient at cooking food: most studies suggest that only 5–10 per cent of the calorific value of the wood goes into heating the pot. In the hilly areas, mainly in Eastern and Southern Africa, and the Sahel, where nights can be cool, the heat lost from the pot is welcome to warm the room. But over most of the continent the wasted heat goes to make sticky tropical nights even hotter.

Improved stoves promise a number of benefits. In laboratory conditions they have efficiencies of 25–38 per cent – in other words, they can cook the same meal with a half or a third of the firewood used by the three-stone fire, with a corresponding reduction in smoke.

It does not follow that they would reduce deforestation by a half or two thirds. Supposing a stove programme were spectacularly successful and reached three out of every four households, enabling each one to cut its fuelwood use by around one third, the country would save a quarter of the fuelwood used for cooking. Assume now that 15 per cent of wood is burned for other purposes, such as blacksmithing or brick-making, then the country's total saving of fuelwood would be only 20 per cent. Even in countries where half of all deforestation was due to overcutting for firewood, a first-rate stove programme would reduce deforestation by only 10 per cent. This reduction would be eaten up by only three years' population growth at current rates. *Hence with fuelwood, a huge reduction in demand can be achieved by reducing population growth.* If Africa's birth rate were cut to the level of South Asia's, the demand for fuelwood forty years hence would be cut by as much as 30 per cent.

Environmental benefits, however, are not the only consideration. For individual households a cut in fuelwood consumption of a third would be a very significant gain. For poor urban households buying their fuel, it would amount to an 8–14 per cent increase in income available for other things. For rural households in semi-arid areas, it would save women four to six hours a week, improve their nutritional status, free extra time for planting and weeding, and boost crop yields accordingly.

All but a handful of stove programmes to date in Africa would have to be classed, to be charitable, as experimental pilot projects, or to be frank, as failures. Many have distributed only a few hundred stoves. Even most of the relatively successful projects can boast only 5,000–8,000 stoves in use in African households. Sometimes the stove design is at fault. All too often backroom boffins develop stoves on the basis of technical efficiency alone. Little research is done on existing stoves or patterns of cooking and fuel use. Models are not extensively (or not at all) consumer-tested before wide release. Many prove unpopular because traditional pots will not fit, because the stoves are too bulky, or because they are no use for room-heating in areas where this is needed. Often they are too expensive for poor households, who are unlikely to pay out scarce cash for a stove as long as they can still collect fuelwood for free. In some cases the stoves crack or crumble quickly, or do not deliver the promised fuel economies.

Sometimes even a good, cheap design fails to take off because the wrong strategies of dissemination have been chosen. A number of stoves are made with cement and need trained masons to erect them. In some instances the concept of popular participation is taken to absurd lengths, and committees are formed in each village to decide what kind of stove people want. Often subsidies are given to reduce the cost to households: but subsidies are not replicable on a national scale, and may enable designs to survive that will flop as soon as they are left to stand on their own feet.

A NEW INDUSTRY IS BORN

The most outstanding stoves programme in Africa is a relative newcomer. Kenya's improved charcoal stoves programme began in 1982 as part of USAID's Kenya Renewable Energy Development Project (KREDP), and did not launch its model until November 1983. The original project goal was to sell or give away 5,000 improved charcoal stoves by the end of 1986 – a modest goal, but a realistic one in the light of past African experience. But by October 1985, it had created a new industry whose major producers alone had sold 110,000 stoves, and revolutionized the market in charcoal stoves.

Four out of five urban households in Kenya cook on charcoal, which is preferred because it is easier to transport than wood, and burns without smoke. The traditional charcoal stove, the jiko, was introduced to Kenya by Indian railroad labourers early in this century. It is basically a small cylinder of scrap metal on legs, with three hinged metal pot-rests, a metal grate pierced with holes, and a flap-door to get the ashes out. Its efficiency is around 19 per cent. It is cheap – around 35 shillings ($2.30) – but the proverb 'buy cheap and you buy twice' applies: the jiko lasts only a year, and the grates have to be replaced every three months or so, at 8 shillings each.

The new jiko began as a twinkle in the eye of a dynamic Kenyan polymath and lecturer in land-use planning, Max Kinyanjui. In 1979 drought lowered the water levels in Kenya's hydroelectric dams, and the country faced repeated power cuts. Kinyanjui went looking for a charcoal stove to cook on. He could find nothing to his liking, so he set out to design his own. The early models were clam-shaped – a hemisphere of metal, lined with ceramic, with a heavy grate and a lid with a built-in chimney. Kinyanjui's friends came, saw and ordered, and within two years he had made more than 700 stoves, continually refining and testing the design on the basis of suggestions and complaints. In 1981 he had a stall at the Renewable Energy Conference in Nairobi. Within a few days he took $100,000 worth of orders, and had to close the books and take his phone off the hook.

When KREDP's Director, Amare Getahun, was looking for someone to head the stove programme, Kinyanjui seemed an unbeatable choice. Kinyanjui and Getahun decided to focus on improving the familiar traditional jiko rather than introducing an entirely new design. The design goals were not only fuel efficiency, but low cost, ease of manufacture, and consumer acceptability. Perhaps the most decisive choice was that they would not attempt to set up new channels of dissemination, but would use the existing network of artisans and dealers who marketed the traditional jiko. These people had the skills, the access to raw materials, and the sales outlets. The jiko was a consumer good, purchased at commercial outlets by individual households. From the beginning Kinyanjui aimed high – his goal was to supplant the old jiko. To do so he would have to use the same channels.

His first design used a traditional jiko outer cladding, lined with a fired clay tube for insulation. But the liner cracked easily, and it was difficult to fit a grate in. In April of 1982, Kinyanjui paid a visit to Thailand which was to prove seminal. There, a ceramic-lined metal charcoal stove, called the Thai bucket after its shape, supported a huge informal industry of artisans. Kinyanjui returned to Kenya, and tried out the Thai liner in the cylindrical Kenyan cladding. The funnel-shaped liner held the grate firmly. It had in-built pot-rests, but these broke easily, so Kinyanjui added hinged metal pot-rests to the metal cladding.

At this point the most crucial phase of the design process began, a phase neglected in so many stove programmes: consumer testing, as extensive, rigorous and thorough as any large Western company would carry out. The tests were organized by KENGO – the Kenya Energy Non-Governmental Organizations Association – and involved 600 households in all. They were given free improved jikos, in exchange for their comments on the product. The trials led to a number of important changes. The cylindrical metal cladding overbalanced when women were stirring pots of thick *ugali* maize porridge. So a waisted design was introduced. The bell-shaped bottom gave the stove much more stability. Only the top half was insulated, with a clay dish with an in-built grate. This reduced the overall weight and cost, and left more air space in the ash chamber.

The final design lights faster and cooks faster than the traditional jiko. It costs two to three times more than the old jiko, but it is an excellent buy simply on grounds of durability: the cladding lasts for two years or more, twice as long as the cheap jiko, and grates last for 18 months, six times as long. Its greatest attraction is fuel economy. The measured efficiency is 29–30 per cent, and most users report fuel savings of up to 50 per cent. For the average Nairobi family spending 170 shillings a month on charcoal *the stove pays for itself in only one month*.

Over a period of a year, it would provide a massive 600–1000 per cent return on the investment, with annual savings equivalent to a month's pay. Outside Nairobi, where charcoal is cheaper and the improved stoves dearer, the period of repayment is two

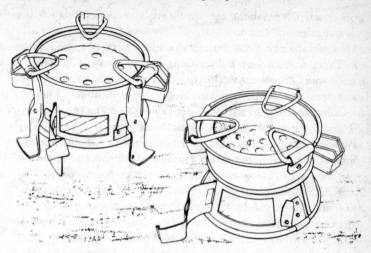

In Kenya the old metal charcoal jiko (left) is giving way to the clay-lined stove (right), with fuel savings of up to 50 per cent

or three months, but the return is still a handsome 300–500 per cent. Emily Wabucheri, for example, who owns a dress shop in Bukura, swears by her improved jiko, which she has had for a year. She used to use five bags of charcoal a month, at 36 shillings a bag. Now she uses less than three. She saved the stove's price, 70 shillings, in less than a month. Gilphers Oduori, who runs a small roadside cafe in Mumias, cooks *ugali* and cowpea samosas on his new jiko. He also has a haybox – a basket of woven grass insulated with straw for keeping food boiling-hot. When he has part-cooked a pot on the jiko, he heats up a flat stone, puts the stone in the bottom of the haybox and the pot on top, and leaves it for two hours to complete the cooking. With the two devices combined, he has cut his charcoal used by two thirds.

The mistake made by previous improved stove projects in Kenya was to assume that appropriate technology needs a special and distinct form of dissemination. In fact many AT devices, from cheap hand maize-grinders to tin tanks to collect rainwater from the roof, have spread successfully in Kenya just as they did in nineteenth-century America, with no aid from government or donors, simply through networks of small businesses, dealers and markets, by word of mouth from satisfied customers, and by demonstration effect among neighbours.

There was already a flourishing jiko industry in Kenya. The improved jiko project decided to harness its skills, resources and contacts. Metalworkers who make the traditional jiko were located, and offered free training in how to make the new stove. The largest concentration of artisans was at Shauri Moyo, Nairobi's popular market, where hundreds of smiths squat among heaps of flattened oil drums and wrecked cars, hammering out pots and pans, painted trunks and kerosene lamps.

KREDP made sure that the new skills did not go rusty for lack of a market: they orderd the 600 stoves for consumer testing from the first trainees at Shauri Moyo. For centres outside Nairobi and Mombasa, they toured markets and trade fairs with a mobile training unit in a van. While artisans were being taught how to make the new jikos, project workers would demonstrate them to visitors. Orders were taken on the spot from interested customers, and immediately passed to the trainees. The beginnings of a local market in improved jikos were created at every stop.

The ceramic liners have to be made separately from the metal cladding, and it is a trickier business. They are kiln-fired from a mixture of clay, sand, dung and water. The project picked out reputable pottery companies, and trained their workers. Seed money was available to cover start-up costs – but it had to be repaid in full from later profits. Making ceramic liners for the new jikos has become big business. The market leader, Jerri International, recently won an order for 25,000 new stoves from the Kenyan army. Army jikos can have a wide-reaching demonstration effect after conscripts return home. Typically for a new industry in a new technology, several Jerri employees have now left to set up their own jiko businesses. Jerri buys metal claddings from the Shauri Moyo blacksmiths, but it will not sell liners to outsiders. To break this potential monopoloy and bring jiko prices down, Max Kinyanjui set up his own ceramic factory, Miaki Jikos, which supplies liners to the artisans.

By 1985, there were around 30 enterprises producing liners or complete jikos, and 98 artisans. The new jikos were on sale from more than 70 retail outlets, from market stalls to food shops and co-operative hardware stores. John Njuguna, a stall-holder in Shauri Moyo, sells 80 new jikos a month, and reports that the old-style jikos are becoming very hard to sell. Peter Muriuki, who makes and sells improved stoves in Chavakali, a small town in western Kenya, says the old model has no market now. Both traders' stock of the traditional jiko moves so slowly that the models on display have gone rusty waiting for buyers.

The project's impact, in just over three years from the initial design experiments, is an impressive achievement in a field littered with failures. Part of the success stems from the design. The technology is labour-intensive. It uses cheap natural materials and recycled waste. It builds on existing metalwork skills, and provides new work for informal-sector artisans instead of putting them out of business. It is cheap, costing less than the old jikos over an 18-month period. Crucially, it repays its costs in a very short period – one to three months – and provides significant savings thereafter. Of course the specific details of the design would not be transferable. They are geared to the appearance of the old jiko, the cooking habits of urban Kenyans, and so on. What can be transferred are the basic criteria for design.

The dissemination strategy, too, was exactly right for Kenya, and the free-market approach would be well suited for portable items of private domestic or farm technology, in countries with a fairly well-developed network of markets and retailers. The approach involves a very significant level of public participation during market research, when the product is tailored as much as possible to match efficient performance with consumers' desires. Then, when the product is marketed, consumers continue to express their influence by buying – or not buying. Subsidies actually reduce the participation of consumers. If people do not shift from an old, inefficient technology to a new, efficient one, without subsidy, then there is something wrong with the cost or design of the new technology or the way it is being disseminated, and project managers can go back to the drawing board.

In 1985, the Kenya Renewable Energy Development Project turned to a new problem. Although the new jiko saved a lot of charcoal, there was a real danger that it might encourage wood users to switch to charcoal. Traditional charcoal production is extremely inefficient, preserving about 12 per cent of the original heat value. In terms of wood use, a highly efficient charcoal stove is actually less efficient than the most basic three-stone fire. Indeed, if the new charcoal stove led to an increase of more than 20 per cent in charcoal users, wood consumption would actually rise, not fall.

It was crucial, then, to put an improved wood stove on the market that would be just as attractive as the new jiko. So Kinyanjui has designed a metal-clad, ceramic-lined model, the Kuni Mbili. It will cost about the same as the improved jiko but it has a higher efficiency – 35–38 per cent – thanks to a deeper burning area. It has the added advantage that it can burn not only wood, but also maize cobs, stalks and other refuse. And it can be quickly adapted to burn charcoal by inserting a second, higher grate and blocking the holes of the lower grate. A door to the burning compartment will allow it to be opened for room-heating. Consumer testing was completed in 1985, and the stove was due for launch in January 1986, but preliminary interest was so strong that 7,000 stoves had been sold by September 1985, simply by word of mouth. It is safe to predict that the Kuni Mbili will be at least as successful as the ceramic jiko. It may do a lot better – indeed it may even completely supplant its charcoal-burning sister.

THE STOVE THAT JILL BUILT

There are situations where the marketing approach is inappropriate. It cannot reach regions or social groups that the market does not reach. There are people with so little spare cash that they cannot afford an outlay of $5, even if it would pay for itself within a month or two. And there are areas where fuelwood is not yet a marketed commodity, and people will not shell out hard money for a fuel-saving stove, even if searching for firewood is a burdensome chore.

For these situations an alternative approach is needed. The most promising designs here seem likely to be simple stoves that resemble traditional fires, that families can make for themselves from cheap or free ingredients. A few such designs are beginning to emerge. One that looks like a winner is the improved mud

The traditional three-stone fire (left) gains massively in efficiency when surrounded with a simple shield of clay, dung and chaff that can be made in half a day (right).

stove developed at the Burkina Institute of Energy in Ouagadougou. The Institute began working on complex models, with two or three pot-holes and a chimney made of cement plastered on a metal frame. These stoves required a skilled mason to construct, and cost around $5. It soon became obvious that they were too expensive and the method of dissemination too slow.

The design eventually perfected by the Institute is essentially an improved, shielded version of the three-stone stove. The materials are all free, and freely available: four buckets of clay, one bucket of millet chaff, and half a bucket of manure, plus water. The mix is left to stand for a week. The stove is built to fit whatever pot the housewife wishes to use it for. Three stones are positioned on the ground, as pot-rests. Around them, a cylinder is built out of the mixture, leaving a small gap around the pot for the smoke to come out. When it has dried a little, a semi-circular hole is cut in the base for the wood to go in. After a week's drying out, the stove is ready to use. Its measured efficiency is around 30 per cent. Theoretically, 50 per cent of wood use would be saved. In use, savings range from between 35 and 70 per cent.

The simplicity of the design allows a simply dissemination strategy. There is no need for skilled masons: anyone can be trained to make their own stove in a half-day session. The actual labour involved totals another half day. It can be done in the early months of the dry season, when women's labour load is lightest. Women (or their husbands) may be unable or unwilling to invest cash in a new stove, but they are only too ready to spend a day's work on a device that will save them as much as twenty days each year in wood gathering. Here again the pay-off period (this time in labour) is extremely short, just a couple of weeks for many women. The costs of materials is nil, the opportunity cost of the labour virtually nil. Villagers have nothing to lose, and much to gain.

The stove is extremely popular with users. The Naam movement (see p. 279) began introducing it around 1981. The villagers of Ouffré, near Ouahigouya in Yatenga province, have had improved stoves for four years. Most of them are still in good condition – any holes that develop are filled in with mud. Many women have two stoves, to fit different sizes of pot. One Ouffré man, Djibriné Ouedraogo, has become an instructor, teaching a score of village women at each session. They bring their cooking pots along; Djibriné shows them how to make the stove. The women make their own under his supervision. Then they break them up, and make them again, and repeat the process. Now experts, they go home and make their own.

The Naam movement is disseminating the model among its member groups. But the improved three-stone stove was still a design looking for a nationwide dissemination strategy. Then, in August 1983, a junior officers' coup in Ouagadougou brought to power a government determined to spread development out into the neglected villages. The government, headed by Captain Thomas Sankara, was deeply committed to halting land degradation, and introduced the 'three struggles' against deforestation, overgrazing and bush fires. Tree cutting was henceforth to be closely controlled. Livestock were no longer allowed to wander around free, but had to be under the supervision of a shepherd, or tied up and fed with cut fodder. Bush fires were banned.

Improved stoves were seen as an important contribution to reducing the pressure on trees and lightening women's burdens. In the three months to January 1984, 1,200 women in Burkina's capital, Ouagadougou, were trained to train other women. By November 1985 there were 30,000 improved three-stone stoves in use in Ouagadougou, and preparations were being made to go national. The mass popular organization, the Union of Women of Burkina, will mobilize women. Each sector of Ouagadougou has been twinned with one of the country's provinces. Trained women from each sector will go out to their twinned province and train around sixty more trainers. These, in turn, will train one or two women from every village in the country, who will train their neighbours. This combination of a pyramidal training structure with popular mobilization was first tried out in the mass vaccination campaign of November–December 1984, and proved that it can bring fast, impressive results at very low cost (p. 267). Progress with the stoves is encouraging: by April 1986, 83,500 improved stoves were in use. It looks a fairly safe bet that, within a year or two, almost every women in Burkina Faso will have built herself an improved three-stone stove.

For all their differences, the Kenya and Burkina programme have a number of common lessons for other stove ventures. They are both ambitious in their goals: the aim in both cases is to reach the vast majority of households in the country. That aim cannot be achieved by dreaming up an exotic stove design in a laboratory: it has to be based on a thorough preliminary study of fuel use, cooking habits, and existing stoves in each country. Wherever possible a widespread existing stove should be used as the basis, and worked on to improve its fuel efficiency – without ever losing sight of the end uses, sizes of cooking pot in common use, types of dish cooked, and subsidiary needs for light and heat where these apply. The cost should be kept as low as possible by using free or cheap local materials. The fuel savings should be enough to repay the cost to the consumer – in cash, labour or materials – within three months. The dissemination method, too, must influence the design: for the market, models have to be portable and easily made by stove artisans. For self-build stoves, the model must be simple and foolproof, so people can learn how to make it in half a day, and can make it in another half-day.

Once the basic design is chosen, there should be very extensive consumer testing in the main cultural regions of a country – in some countries a number of different models may be needed to suit different regions. The design should then be reworked, taking into account consumer problems, and re-tested on a sample of households before launching.

The market approach is ideal for nationwide dissemination in countries like Kenya, Nigeria or Ivory Coast, with decent communications and thriving rural markets reached by most types of goods. Frequently the private market can and does spread appropriate technologies quite unaided – from cycle rickshaws in Asia, to tyre sandals in Africa or pots and pans made of old tin cans in the Andes. But where real gaps remain, there is a role for non-profit bodies, from governments to voluntary agencies. They can organize the research and development that may be too expensive for small businesses. They can provide risk capital, and promotion in the public media.

In all cases they should enlist existing artisans to help design the new technology, and train them to use it, so they are not put out of work, and use existing channels of wholesalers and traders.

The self-build approach is better for the poorest countries, where people have very little cash and rural markets are spartan. Any and every channel can be used for dissemination, from women's groups to schools, but few channels are as effective and speedy as Burkina's pyramid-training approach.

Most countries will fall in between these two stools, and may well need both kinds of programmes: the market approach for cities and towns, the self-build approach for rural areas and the poorest squatter settlements.

13 Pastoralists and ploughmen: livestock

Nigeria, Africa's most populous country, straddles most of the continent's ecological zones, from rainforest in the south to semi-arid savannah in the north. In the south the only local livestock are goats, miniature specimens not much bigger than kids. They roam between the rust-roofed houses, scavenging in open gutters, scattering from the roads as 'mammy wagons' overloaded with people or yams tear through at homicidal speeds. Nigeria's cattle live in the north, herded by pastoral tribes like the Fulani. You see their dry-season huts in the harvested fields, tents of straw looking for all the world like haystacks. Right across Nigeria's infertile middle belt, you can sometimes spot cone-hatted herders with white shifts, driving scrawny herds of long-horned cattle through crackling, desiccated bush, on their way south. Most go by rail these days, heading for the crushed cattle markets of Ibadan and Lagos and the slaughterhouse. Vast humid areas without cattle or horses, vast semi-arid areas where cattle-herding is a specialized business – Nigeria's paradoxes encapsulate those of a whole continent.

Almost everywhere in Africa, livestock are a resource that falls far short of its potential. And in the marginal farming areas, they pose a severe threat to the environment. Livestock figures more prominently in Africa than in any other region. Because of her vast drylands, Africa has the largest extent of permanent pasture of any continent – 778 million hectares in 1983, an area four times larger than her cultivated cropland. The continent has the largest number of pastoralists – somewhere between 15 million and 25 million people depend primarily on livestock, with more than a million each in Sudan, Somalia, Chad, Ethiopia, Kenya, Mali and Mauritania. Livestock is estimated to be worth about one third of the value of total agricultural production.

In most parts of the world, livestock have played a pivotal role in the development of farming. Thanks to their peculiar digestive system, ruminants are able to convert into useful products resources that would otherwise be wasted, from crop residues to the pastures of lands too marginal for agriculture. They provide protein in the shape of milk and meat. The larger animals serve as a source of power in transport and ploughing – it has been estimated that a family can cultivate two or three times more land with a pair of oxen than with hoes. Livestock provide high-quality fertilizer – the manure from a single cow can boost low-level grain yields on a hectare of land by 25 per cent. Their fodder needs allow new crop rotations, including legumes, that can maintain soil fertility with shorter fallow periods.

But Africa's peculiar biogeography blocks the full promise of livestock over most of the continent. The reason is the tsetse fly, carrier of the potentially fatal livestock disease, trypanosomiasis. The fly infests around 10 million square kilometres, covering almost four fifths of the humid and sub-humid zones, and half the semi-arid area. Tsetse effectively severs farming and livestock rearing over most of the continent, to the detriment of both crops and animals. (See map, p. 350.)

East Africa is the only region where the two activities are properly integrated. Because of the cooler temperatures, only a fifth of the highlands are affected by tsetse. 17 per cent of tropical Africa's livestock live here on only 4 per cent of her land.

The best potential for fodder production lies in the humid areas, 18 per cent of the land area. But only 6 per cent of Africa's livestock are found here – scattered pockets of dwarf cattle, sheep and goats resistant to trypanosomiasis. In this region, humans – chiefly women – are the main beasts of burden; the hoe stands in for the plough, and the head for the horse's back.

The major concentration of livestock, however, is in the semi-arid and arid areas which have 58 per cent of the stock. But low rainfall means low potential for fodder. Settled farmers in these regions do own livestock, but in West Africa most entrust their cattle in the rainy season to specialized herders who drive them north to seasonal pastures on the desert fringes. In the dry season the herds return and graze on the stubble of harvested fields. Farmers have their reasons for this seasonal separation from their cattle. In the rains the tsetse danger is greater in the home villages, and the cattle might eat or trample crops, cause disputes, or divert labour from working the fields to guarding the herds. Out on the ranges they profit from free pasture and the nomads do the guarding. But farmers pay a high price. They lose much of the draught power, milk and manure of their beasts. The benefits of integration are lost.

LIFE ON THE MARGINS

The pastoral nomads face a different set of problems. Africa's rainbelts are mobile, drifting north and south with the seasons, bringing water to parched lands and colouring the desert margins with flushes of green. The nomads who exploit these shifting resources must be mobile too, often over distances of hundreds of kilometres. Their mobility is their principal defence against the fluctuations of Africa's climate, her periodic droughts and her uneven, spotty rainfall.

Africa's nomads, like her farmers, are the world's experts in mastering a complex and contrary environment, in harnessing marginal and constantly moving resources to support human life. The rains in the rangelands vary more than anywhere else – by 40–60 per cent up or down on the trend from year to year. Pastoralists have to be masters of insurance, and they take out multiple policies. Many tribes mix cattle and camels, goats and sheep in their herds, just as farmers mix millet and sorghum. If the rains are good, the cattle and sheep will do well. If they are bad, the hardier camels and goats will survive. Herds are split up among relatives – or as loans to poorer families – so one family's animals are spread widely across the pastures, and cannot all perish on one dry patch. Where they can, owners build up their livestock numbers, so some will pull through in a drought. 'You don't need 300 cattle,' livestock expert Jean-Claude Bille protested to one herder in the Sahel. 'If there is a drought, you'll lose 250.' 'Yes,' replied the wily nomad, 'but I'll be left with 50 and I'll survive. If I had only 100 cattle, I'd be left with only twenty, and with twenty I'd die.'

The most critical problems of the pastoralists are to some extent endemic in the precarious ecological niche they have chosen. The long dry season takes a heavy toll of livestock and humans alike. The pastures and ponds dry up, the distance between watering points increases, the animals face long walks and long intervals between waterings. Milk production plummets. In the wet season, 2–3 litres of milk a day can be drawn from each cow. By the end of the dry season, there will be no surplus at all over the needs of calves. Pastoral people, as well as their animals, suffer from heat stress, lack of water and shortage of food. By the time the next rains arrive, adults may have lost 5 per cent of their weight, and children are weak and vulnerable to disease. Animal diseases are rife: blackleg, brucellosis, foot and mouth, rinderpest. Not surprisingly, the mortality of calves is high, averaging 36 per cent in Niger for example, a massive waste of potential. Human child mortality is probably not much less.

Drastic rainfall variations from year to year mean drastic changes in the ability of the ranges to support animals. The pastoralists were always able to absorb minor and medium-scale fluctuations but major droughts, lasting more than a year, would always have been an intractable problem. The only sensible response was flight. Even then many families would lose most of their herds, or become involved in territorial disputes with other clans and tribes whose land they moved into. Stock losses in recent droughts give some indication of the risk. In the Marsabit area of Kenya, 20–30 per cent of all cattle died in the 1971 drought, and the same proportion again a decade later. In the region north of Tahoua, in Niger, two thirds of the cattle and one third of the goats died in the 1973 drought. Eleven years later, when numbers had still not recovered to their previous level, the 1984 drought killed three fifths of the herds.

Pastoralists often get the blame for desertification. The conventional view has it that they overstock and overgraze. Excess cattle shift the balance of vegetation from deep-rooted perennials to annuals, which are less able to protect the soil against erosion. Their hooves compact the soil and make it less permeable to rain. To get an extra flush of green vegetation, pastoralists set fire to the range, wasting millions of tonnes of potential forage, plus nitrogen and other nutrients essential for a healthy soil.

But the consensus is now shifting. It is now widely accepted

that the ranges are surprisingly resilient, and re-establish themselves quickly if they are left alone for a year with decent rain –
as happens automatically when a drought kills half the animals.
Fire and grazing have undoubtedly changed the vegetation of
the pastureland over the centuries. They have reduced the
diversity, but they have probably increased its ability to support
livestock and human life. Grazing, for example, reduces the loss
of moisture from the soil. Without grazing, the soil is dry within
two weeks of the end of the rains: with grazing, plant growth is
prolonged for two months.

Before colonial times, many of the pastoral peoples compensated for the poverty of their natural habitat by extracting
surplus from settled farmers: their harsh environment gave them
hardiness, their cattle mobility, their horses and camels speed
and power. They dominated the trade across the Sahara desert,
and in many areas founded conquering dynasties like the Fulani
emirates of Northern Nigeria.

Since colonial days the proud nomads have suffered a secular
decline. Farmers acquired firearms. Colonial governments shifted
the direction of trade away from the Sahara towards the coasts,
and set up regimes that related more easily to settled cultivators
than to shifting herders. The successor states, too, were mostly
dominated by farming tribes. The pastoralists were economically
and politically marginalized, and largely by-passed by modern
health and education services.

Other changes worked to reduce their range and their mobility.
In the Sahel the wet 1950s and 1960s shifted the limits of
crop cultivation and farmers moved north. The nomads moved
northwards, too, as what was once desert scrub had become
potential pasture. The dry period that set in from 1968 moved
the vegetation belts south again, but most of the farmers stayed
put and the pastoralists could not so easily move back south.
The herders were sandwiched between desiccation moving down
from the north and the farming frontier moving up from the
south. In East Africa too, the rangeland fringes were frayed
away by farmers moving down from the overcrowded hills.

The scene was set for a situation where pastoralists and their
herds could do real and serious ecological damage, not so much
in the ranges themselves as in the farming areas where many

pastoralists spend the dry season. Many of these areas are marginal and unsuitable for long-term cultivation. Cultivation itself reduces the cover of vegetation, but in the dry season local farmers' stock, expanded by the nomads' herds, remove every last vestige of vegetation within reach. There is no doubt that the combined impact of local and nomad stock contributes to crusting, destabilization of sand dunes, weakening of tree cover and erosion. Settled farmers might be able to control their own livestock numbers to reduce the damage, but they have no power to control the nomads' herds, or to exclude them from areas where they have a traditional right to graze.

A PICTURE OF ALMOST UNRELIEVED FAILURE

No sector in Africa needs effective help more than livestock. No sector has had less. Livestock projects have probably the worst failure rate of all in Africa. In 1985 a confidential World Bank report surveyed the record of 330 projects partly or wholly concerned with livestock components. Economic rates of return ranged from 6 per cent in East Asia and 16 per cent in South Asia. In East Africa, pre-project assessments predicted average rates of return of a healthy 13.3 per cent. In the event, the projects turned in *losses* averaging 1 per cent a year. In West Africa, the results were worse still: profits of over 19 per cent were predicted; instead, the average project lost 3 per cent a year.

Within this landscape of failure, projects to help pastoral groups have been a particular blot. Reviewing two decades of experience, US anthropologist Walter Goldschmidt commented: 'The picture that emerges is one of almost unrelieved failure. Nothing seems to work, few pastoral people's lives have improved, there is no evidence of increased production of milk and meat, the land continues to deteriorate, and millions of dollars have been spent.'

The failure is not surprising. Even given the most talented, open-minded project directors with unlimited resources, the job of helping pastoralists would have been a hard one. The areas involved are vast and inaccessible. Reliable data is almost impossible to collect because of climatic fluctuations: an area that appears an empty desert at one visit may be a lush pasture a month or a year later. The ecology of the ranges is complex and diverse, as are the cultures of pastoral groups, and both are only just beginning to be sympathetically understood. Until recently most projects did little to correct the general ignorance: three out of every five World Bank livestock projects between 1967 and 1981 used no anthropological input at any point in the project cycle. Nor were pastoralists in a position to correct misconceptions. They had little political influence and few contacts with local administrators. Projects involved only the most fleeting consultations, more usually none at all.

In this vacuum of accurate information, it was easy for the usual pro-Western cultural bias to thrive. Western advisors and African governments concurred that African pastoralists were (along with African farmers) hopelessly inefficient and destructive, and should be transformed as swiftly as possible into commercial ranchers just like their American or Australian counterparts.

Meat production was low, while in the cities, and in the tsetse-infested areas, millions of consumers could not get hold of cheap, good-quality meat. Projects put the emphasis on the production of meat for sale, and built huge marshalling yards and abattoirs. But the pastoralists' prime concern was milk for their own consumption. When veterinary services increased their herds, they kept most of the increase instead of selling it. The yards and abattoirs worked at a fraction of capacity.

Desertification was the other major concern. The essential problem here, according to the original Western analysis, was that the ranges were communally owned, while the herds belonged to individual families. The 'tragedy of the commons' was playing towards its final act. No one had an incentive to conserve or improve the rangeland, since they would bear the whole cost but enjoy only a small share of the benefit. On the contrary, each individual owner had an incentive to overstock, since he would reap the full benefit of each extra animal, but pay only a fraction of the ecological costs in range degradation.

The answer, so it seemed, was to privatize ownership of the rangeland. In some cases, as in Uganda, vast areas of communal land were handed over to individual owners. In other areas, as in Kenya's Rift Valley, limited groups of nomads were allocated full title to an area, in the hope that they would begin to see the light and balance their herd numbers with the carrying capacity of the land. But ranching projects had disappointing results: their performance in output was little or no better than that of traditional systems. Pastoralist ranchers were still exposed to drought, and continued to overstock as an insurance. Indeed, in drought years, if the whole ranch area was affected, they would leave the area altogether in search of greener pastures. No ranch could be big enough to provide a permanent escape from drought.

The 'tragedy of the commons' argument was always over-stated. In pre-colonial days, pastoralist clans or tribes did have some power to regulate access to the pastures, mainly through their control over dry-season wells. That power was taken away from them when colonial and independent authorities nationalized the ranges, and sank boreholes which were thrown open to all comers. The area around the deep wells and boreholes became heavily overloaded. Circles of 8–12 kilometres' radius around each one – the distance a cow can range between waterings – were degraded and deforested.

So new controls were brought in, in an attempt to balance herd numbers with the state of the range. In some cases the right to use the range, or ranch, was made conditional on following the instructions of range managers who had far less information about local conditions than the pastoralists themselves. As British livestock expert Stephen Sandford has written: ' "Paddocks" of up to 1,000 square kilometres, with square, straight line boundaries, are opened and closed to all species of livestock simultaneously according to a rigid schedule sometimes laid down years in advance.'

Once again, the wayward nature of the African climate had been ignored. The pastoralists followed the wandering rains, as they must if they were to survive. Where they were not allowed to, they were pauperized.

FIRST STEPS

The pastoral problem in semi-arid Africa is the most intractable we have considered. There are no easy solutions, no spectacular success stories of the kind we have seen in agriculture. But there are promising beginnings, and a new humility, an openness to learning, a readiness to experiment on a small scale, which may well pay off over the next five years or so.

The old arrogance towards the African herdsman has disappeared: most people now accept that pastoralists in Africa are a great deal more efficient than outsiders ever realized. They may not sell as much meat as city dwellers and governments might like, but their goals are different. Studies in southern Ethiopia have found that the Borana tribe manage to extract four times as much protein per hectare, and six times as much food energy, as commercial ranches in Australia's Northern Territory – and support six to seven people per square kilometre of rangeland, against one person per 500 square kilometres in the Australian ranches.

Any improvements to this record are bound to be marginal. The first imperative is to respect the pastoralists' prime concern with milk rather than meat, and with security before cash income. One of the most promising lines lies in improving on the high mortality rate and low weaning weight among calves, in ways that do not reduce the availability of milk for humans. Better veterinary services are one approach that pastoralists themselves are enthusiastic about: they cut down the waste of resources that death and disease represent. Projects among the Fulani, Tuareg and Maasai use the animal equivalent of barefoot doctors. These 'paravets' are herders trained to do basic injections and cures so that veterinary services are within reach of the most remote groups. In its Ethiopian Rangelands programme, the International Livestock Centre for Africa is exploring, together with Borana pastoralists, ways of improving calf nutrition without diverting cows' milk from humans. Many of the Borana have taken up farming in response to drought. ILCA researchers are testing the feasibility of growing fodder legumes as part of a crop rotation. Combined with maize stalks, these can reduce calf mortality and improve weaning weight. More male calves survive at a healthy weight, so more can be sold without reducing the overall size of the herd. The nitrogen fixed in the soil by the fodder legume, meanwhile, boosts the yield of

associated food crops as well. ILCA's sub-humid zone programme around Kaduna in northern Nigeria is experimenting
with fodder banks – reserves of leguminous perennials like the
low, bushy *Stylosanthes*, which can be used in the dry season to
improve calf nutrition.

Few experts still entertain illusions about balancing stock
numbers and the carrying capacity of the rangelands, at any
scale. Remote sensing by satellite can give a general idea of the
state of pastures and is useful as an early warning, so preparations for drought relief can be made. But there is no easy way
of getting detailed information to pastoralists on the ground
that would be any quicker than their own bush-telegraph. No
approach that involved limiting the pastoralists' mobility would
stand any chance of acceptance, as they depend on mobility for
survival. No proposals to cut herd numbers would be accepted,
as large herds are a drought insurance. In any case, after a
succession of bad years, pastoralists' herds in many parts of
Africa have been reduced to the minimum needed for survival –
and often below. Among the WoDaabe of Niger, only one family
in ten had herds big enough to support them in 1981–2 – and
that was *before* the most recent drought.

If there is any man-made degradation due to overloading of
the rangelands, it is happening wherever animals are artificially
concentrated – around dry-season boreholes, trading centres
and, increasingly, famine relief camps. Beyond these limited
areas there are often vast tracts of rangeland underused through
lack of water points in the dry season. Only 16–21 per cent of
the Marsabit area of Kenya was overstocked in the early 1980s,
according to Walter Lusigi, director of the Integrated Project in
Arid Lands. The project area was capable of sustaining double
the actual livestock population. The solution to reducing degradation lay not in reducing livestock numbers by enforced sales,
but in improving the *distribution* of livestock across the area. This
could be achieved by increasing the number of watering points,
providing a well every 12 kilometres or so. ILCA's Ethiopian
Rangelands Programme is helping Borana nomads to increase
the number of ponds, where water remains for 4–5 months into
the dry season. More watering points would also improve the
state of the herds and of their owners during the dry season.

The other approach to balancing stock numbers and range resources is to try to restore to pastoralists the control over access that they used to enjoy in pre-colonial days – for example, by encouraging the spread of hand-dug wells, and giving those who dig them control over their use. A number of projects in Senegal, Niger and elsewhere have attempted to end the pastoralists' marginalization by setting up pastoral associations based largely on existing kin groups who share dry-season pastures. The associations express pastoralists' needs to government and channel government services to pastoralists. Credit programmes, administered through these associations, can help to strengthen bargaining ability with farmers and traders. Nomads can buy grain and store it when it is cheap, just after harvest, and delay selling animals until prices are favourable.

IMPROVING SECURITY

Pastoralists, as we have seen, have developed very sophisticated systems of insurance against drought. In a major drought covering a vast area, all these insurances fail. Drastic stock losses occur, the markets are flooded with scrawny animals, and prices plummet. Drought devalues the pastoralists' capital. Proud, independent nomads become paid herdsmen, night guards, prostitutes, beggars or refugees in relief camps.

Most livestock experts concede that there is no satisfactory answer to the problem of major droughts among pastoralists that poor countries could afford. In developed countries with arid rangelands, like Australia and North America, national insurance schemes can cover ranchers against catastrophic losses. Countries with ample slaughter and storage facilities and a ready export market can buy up livestock at reasonable prices at the beginning of a drought – though there is a real problem here in deciding when to act, as rains could arrive belatedly at any time. But abattoirs and deep-freeze facilities big enough to cope with the flood of animals that appear during droughts would be at least three times too big for non-drought years, and few African countries could sustain the excess capacity.

The inescapable fact is that the resource base of pastoralists is subject to wild fluctuations. It would be safest to limit stock to the carrying capacity of the driest years, but that involves forfeiting a great deal of production in wetter years, and is not possible in Africa without depriving millions of pastoralists of their livelihood. On the other hand, if, as happens at present, you build up stock to the limit that can be supported in the wettest years, you face the certainty of drastic livestock losses in dry years resulting in death and dispossession among nomads. Nor can the balance be restored quickly when the rains return. Cattle herds expand at around 11 per cent a year even in favourable circumstances, and would take four years to recover from a one-third drop, and eleven years to recover from a two-thirds drop.

More modest efforts to cushion the impact of fluctuations are possible. In Mali, surplus livestock are processed cheaply into dried meat, though there is a limit to what the market will stand. If pastoralists are paid a reasonable price for their animals, they can avoid pauperization and acquire funds with which they can buy animals after the drought is over to build up their herds again. In the Turkana and Isiolo regions of Kenya, the British charity Oxfam introduced programmes to restock dispossessed pastoralists with herds of goats and sheep, which build up in numbers faster than cattle. Before and after a major drought, destocking and restocking programmes on the scale required would be far beyond the capacity of government budgets already strained by food and cash crop shortfalls. Until African countries become rich enough to sustain drought insurance schemes of their own, the international community should provide the insurance, financing destocking and restocking programmes on a pre-planned basis, so pastoral communities can be sustained in their independence, not assembled as destitutes in relief camps.

Perhaps the most promising approach to improving security involves diversifying the resources on which pastoralists depend. Encouraging wider use of camels and goats is one avenue. Both can go longer without water than cattle, so they range more widely around water points and do less ecological damage. Both are more likely to survive droughts than cattle, and form a nucleus from which a bigger herd can be reconstituted. Goats reproduce much more quickly, and can recover from a one-third drop in a year and a half, and from a two-thirds drop in three and a half years. Camels can provide two and a half times as much milk per hectare as cattle, and goats two and a half times as much meat. Camels, goats, sheep and cattle all have different eating habits, so mixed herds can achieve more useful output from a given area than herds of a single species.

Reserves of fodder trees, where pastoralists could come in dry season or drought, are another possibility. The Niger Integrated Livestock Project at Tahoua is exploring the idea of giving small groups of nomads exclusive title to areas where drought-resistant trees or shrubs like *Maerua crassiflora* would be planted. *Maerua* is deep-rooted, and is still growing actively eight to ten months after the last rains. It is prolific, nutritious, and palatable. Its leaves and fruit are eaten by nomads themselves in drought, its bark and leaves are used against fevers and toothache, and its tough wood for weapons and tools.

The final form of diversification involves reducing nomads' dependence on their precarious pastoral existence by encouraging (not enforcing) partial sedentarization, helping nomads to acquire a settled agricultural base from which they could send out their herds in the wet season. There are problems here, too: the shortage of promising sites, the danger of desertification around the home base. But drought is driving nomads into farming anyway. Once they have been taught the basics, pastoralists are well placed to integrate their livestock more closely with farming. They can serve as an example of mixed farming to settled cultivators.

In the long run, the trickle away from pastoralism will continue, becoming a flood in drought years. In the meantime, it *is* necessary, and possible, to make worthwhile improvements in pastoralists' lives. What has emerged to date is a method rather than a solution, a way of asking the right questions rather than providing a set of answers. The best projects respect the pastoralists' unrivalled expertise and efficiency within their own difficult environments. They build on that expertise, and co-operate with pastoralists to develop solutions that are acceptable and will work.

LEARNING TO LIVE WITH THE FLY

The greatest challenge to livestock in Africa is the tsetse fly and the parasitic disease, trypanosomiasis, which it carries. The major impact is to ban mixed agriculture from an extent of 10 million square kilometres, an area with the best potential for fodder production in Africa. It has been estimated that this vast area could support 100–140 million extra cattle and as many sheep and goats, and generate up to $50 billion a year in meat, milk, and additional food production.

The tsetse fly is a formidable enemy to contain. Most species of tsetse require a retreat of bush. Bush clearance has been tried as a line of attack – but re-invasion from neighbouring areas is always on the cards. Residual insecticides like DDT can be used, but the economic cost is high – between $400 and $2000 per square kilometre – and the ecological impact dubious. Nigeria has cleared a massive 200,000 square kilometres by spraying, but most countries cannot afford the expense. The bill for clearing the whole tsetse area in Africa would come to at least $4,000 billion – more than the annual gross national product of the United States. Biological methods such as the mass release of sterile males would not come any cheaper.

In recent years, the fly has made advances, taking back some 21,000 square kilometres in Cameroon, 11,700 square kilometres in Zambia, vast areas of Angola, and re-invading Zimbabwe from Mozambique. Work is advancing on lower-cost approaches to tsetse control. The most promising seem to be traps, baited with attractants that smell like cattle breath, dyed in the tsetse's favourite colour. These have been tried with some success in Burkina Faso and Zimbabwe, but a mass dissemination approach has yet to be perfected.

If the enemy cannot be repulsed, its targets – livestock – can be protected. Despite long years of research, little progress has been made in developing a vaccine against trypanosomiasis. The problem here is that the elusive trypanosome has the disconcerting ability to switch around the proteins on its surface and evade the defences of hosts' immune systems. A number of long-standing drugs exist that kill the parasite after an infection has occurred. If cattle are regularly dosed, they can reach milk and meat output levels in the tsetse zone only 20 per cent below their yields in tsetse-free areas. Drug treatment is economical in zones of light tsetse challenge – but it has hardly spread outside the ranching sector. Only 25 million doses of drugs are given each year – one tenth of the number needed to protect exposed cattle.

Drugs involve cash investments, sophisticated management and imports, so they are not a promising avenue for Africa's smallholders for the foreseeable future. The best prospect lies in expanding the stocks of animals that are resistant to trypanosomiasis. These number an estimated 7–8 million cattle, 11–12 million sheep and 15 million goats. They are mainly dwarf breeds. The dwarf N'Dama cattle are strong enough for draught work, which will be mainly transport or pulling of equipment for minimum tillage – ploughing in humid areas is not advisable (p. 135) except on heavy clay soils. N'Dama productivity compares well with that of the larger, non-resistant Zebu outside the tsetse area. Breeding stock imported into Zaire have been multiplied on large ranches, then loaned out to smallholders, who repay the loan after five years in the form of animals. Zaire, which once had no resistant cattle, now has a quarter of a million head. But the spread will be a slow business. To multiply the existing resistant herd to the 100 million potential would take at least thirty years, even with free mobility across frontiers. High-tech approaches could speed the process. In 1983, N'Dama embryos from the Gambia were deep frozen and flown to the International Laboratory for Research on Animal Diseases in Kenya. There they were transplanted into the wombs of sturdy Boran heifers, and by April 1984 ten calves had been born.

In the meantime, the approach for the humid areas must be to work with what exists: to help smallholders to improve the productivity of sheep and goats, which are capable of providing more than twice as much meat from a given input as cattle. These animals can provide the basis of a new form of intensified mixed farming, different from the European version based on cattle. In this version, crops would be grown in alleys (p. 192) between hedges of fodder trees like *Leucaena* and *Gliricidia*, regularly pruned for their leaves and poles. The alleys could be rested by growing ground-level leguminous fodder crops. Small stock would be fed with the leaves in enclosures, rather than left to roam free, and the manure transferred to the fields. The draught power that oxen supply would be gradually provided by improved hand tools and mechanization, while the need for hoeing or ploughing would be reduced by no-plough methods. The tsetse has so far spared most of the fragile soils in humid areas from the plough: this window of opportunity should be used to introduce minimum tillage (p. 192).

THE FAVOURED HIGHLANDS

The highlands of East Africa have the best prospects for livestock development in the continent. Their cooler temperatures put them out of reach of the tsetse fly, so crop farmers can keep their cattle at home. As population density has increased, livestock and arable farming have been gradually integrated to the benefit of both.

Perhaps the clearest success story in the whole African live-stock sector is the development of dairy farming in Kenya, based on crosses of local cows with European breeds to produce cross-bred 'grade' cows that yield up to six times more milk than local breeds. In 1960 there were only 80,000 'grade' cows on smallholdings in Kenya: fifteen years later there were 550,000.

The factors behind this astonishing expansion – averaging 14 per cent a year – were not unlike those at work for Zimbabwe's maize: favourable prices, good market access, good supply and extension services. Between 1960 and 1977 the price of milk rose more than twice as fast as the price of maize. Ready access to the main processor, Kenya Co-operative Creameries, was assured by a network of 300 smallholders' co-operatives, with their own delivery trucks taking the milk from collection points, using improved dirt roads. The government provided the back-up of extension advice and an artificial insemination service that in 1980 made more than half a million shots.

Kenyan farmers have paralleled these government-inspired initiatives with their own. As land has grown increasingly scarce, cattle in many parts of the highlands are no longer grazed freely. Most are permanently tethered on a small grassed patch on the farm, or even kept in stalls and fed with fodder and wastes. This 'zero-grazing' system involves more labour than free-roaming – but it has massive advantages. It reduces the pressure of grazing on plant cover, and of hooves on soil. It allows perennial high-yielding fodder plants like Napier grass to be grown. It greatly increases the availability of manure, which is concentrated in one spot rather than scattered all over the district. The main attraction is that it raises a cow's milk output by anything from two to six times by reducing the amount of energy the cow loses in moving around.

The largest livestock concentration in Africa is in Ethiopia. Wherever you look there are long trails of mixed herds of cattle, goats, sheep, horses and mules, ambling across fields as far as the eye can see. The country has some 26 million cattle, 23 million sheep, 17 million goats, 3.9 million donkeys, 3 million horses and mules, and 1 million camels. It is the natural location for the International Livestock Centre for Africa, with headquarters in Addis Ababa.

ILCA's Ethiopian Highlands programme has been working on ways to improve the use of livestock and their integration with farming. One of the main problems ILCA has identified is that most farmers, paradoxically, do not own enough cattle. Only one third of Ethiopian farmers possess the minimum of two oxen needed for conventional ploughing. Those with no oxen have to hire. Those with one ox mutually exchange with a neighbour, but this often means that planting is delayed, and yields may be cut. In 1983, ILCA developed a new plough design that allows a single ox to plough. It is based on a modification of the traditional *maresha* plough. The *maresha* consists of a long wooden beam pulling a forward-pointing plough-share: the force of the blade cutting the soil pulls the beam downwards, pressing on the double yoke across the ox's neck. The one-ox plough has a metal skid underneath the beam which takes the brunt of this force. Using the one-ox plough a single ox, properly fed, can cover 60–70 per cent of the area covered by a pair. The cost of the adaptation is minimal. The farmers themselves, who maintain their own ploughs, can make the modification.

The one-ox plough has spread rapidly from the 27 farmers who first volunteered to try it out. More than a hundred of their neighbours came forward to request a plough. Another 600 farmers from Dinki, on the edge of the central escarpment, lost all their livestock in the drought and have been given oxen, seed and one-ox ploughs so they can start cultivation again.

With the ox-drawn pond scoop, cattle can be used to improve water
supplies

ILCA has developed a number of other low-cost tools to make
better use of animal power. Watering livestock is a major
problem in the dry season: farmers may have to break off
ploughing at midday to take their oxen to the nearest well or
spring. ILCA has developed a sturdy ox-drawn iron scoop that
can dig out large ponds to store water for the dry season. Special
ILCA-designed mould boards can be added to the traditional
plough so it can be used for rapid terrace-building, or for making
raised beds to allow crops to be grown on fertile but usually
waterlogged dark clay soils. ILCA is also working on ultra-low-
cost biogas digesters. Dung is burned as fuel in most of Ethiopia
so its value as manure is lost. Biogas plants digest it to produce
methane gas for domestic use – and leave a rich slurry that can
be used as high-grade fertilizer. ILCA digesters are made from a
sausage-shaped sleeve of PVC mixed with old engine oil and
bauxite dust to protect it against ultraviolet. The digesters cost
only $40–$50 each. The cost would be brought down as low as
$20 if polythene were used.

ILCA's innovations are having a strong impact on peasants

close to its out-stations. The farmers of the Kormargeffia Peasant Association, close to the station at Debre Berhan in the central highlands, were intrigued as they saw the first pond dug out with ox-scoops. They waited and watched to see if it would hold water. When they saw that it did, they came and asked ILCA's help to make one of their own. They were given the loan of 18 scoops and worked rotas. Every day eighteen farmers turned out, with one pair of oxen each, and dug out a huge pond 70 metres square. Before they had the pond, farmers lost half of every ploughing day walking their oxen to water. The pond has given them an extra two or three hours' ploughing a day.

Many of the members have adopted other innovations. Tsegay is a painfully thin but cheerful 23-year-old. He came to Kormargeffia ten years ago from drought-stricken Wollo province. At first he worked for his keep as a labourer with a local family. When he married, the Peasants' Association gave him land of his own, in five small parcels totalling around a hectare. He grew barley, but even in a good year he reaped a meagre 0.6 tonnes, less than his family needed to survive, and had to work as a labourer to earn cash to buy more grain. Tsegay had a single ox, and had a swapping arrangement with a friend. 'When I loaned my ox to him,' he remembers, 'I was sitting doing nothing till he finished. When it was my turn to borrow his, it might be a holiday [the Ethiopian Orthodox Church forbids work on 160 days of the year] and I might miss one rain. Even if I am a single day late, I lose yield.' In 1983 he heard that ILCA was looking for volunteers to try the one-ox plough. He was sceptical at first – like his neighbours, he thought that ILCA had special extra-strong cattle, or was giving them special feed. He thought his own ox would be too weak – until they let him try the plough with his own animal. He was convinced, and took one of the ploughs. From then on he could plough his own land without delay.

One good experience with an innovation often opens a farmer to others. The same year Tsegay heard that ILCA were selling high-yielding cross-bred cows, three quarters Friesian, one quarter hardy Boran, for $200 each. He sold all his livestock – two calves and five sheep – to raise the cash, keeping only the draught ox. The grade cow now has a cobbled stall, penned off with poles, inside Tsegay's hut. This single cow has transformed Tsegay's life. She gives an average of 5–6 litres a day for nine months of the year – five or six times as much as the local breed. The milk is sold at $0.25 a litre to a van that passes every morning. The annual income from the milk alone is $375. The cow has just had two calves, and Tsegay sold one for $235. Altogether, the cow brings in an average of around $450 a year. Tsegay's cash income has risen fivefold. He has been able to rebuild his herd, now two oxen, a heifer and ten sheep. He has built a bigger dry-stone house for his family. He took a massive risk at the outset, investing half his capital – but the return on the investment has been over 200 per cent a year.

ILCA is essentially a research institution. It has no mass extension arm, and its innovations have not spread far as yet. Their true value can only really be assessed when they pass or fail the acid test of wide dissemination. Perhaps ILCA's most significant contribution is not any specific innovation but rather a way of tackling problems that can bring results in Africa. It views livestock not as an isolated sector, but always as part of a farming system that includes crops and often trees and shrubs as well. It works not just to improve people's livestock production, but their whole output, by integrating livestock with agriculture and agroforestry. It studies existing systems carefully before intervening, building on their strong points and strengthening their weak points. It works together with small farmers and pastoralists, consulting them at every stage. The resulting approaches are low-cost; with low or nil import content, they are culturally and economically acceptable and can be locally managed. Of course, the depth and professionalism of ILCA's studies come too expensive to be replicable across Africa: participation is a cheaper way of ensuring that innovations correspond to farmers' needs.

Outside the highlands, we can expect no miracles in Africa's livestock sector: the problems are too deep-rooted for that. But programmes based on these firm principles are guaranteed to make steady progress.

14 Tackling runaway populations

Only lowland and tsetse-infested bush separates Tongogara in Zimbabwe from the Mozambique border. It is one of the resettlement schemes on repurchased former white farms, set up to relieve population pressure in the overcrowded communal lands.

Tongogara will soon be overcrowded itself. Like the other settlers, Samuel Chiodza was given 12 acres of land when he arrived – double his holding in the area he came from. When I think of Africa's booming populations, I shall think of Chiodza, a bald-shaven 45-year-old in ragged trousers and green sweater. Chiodza has four wives and twenty-eight children. He is aiming for fifty before he has done. 'When you have many children,' he explains, 'they can all get different jobs. One can be a builder and mend your house, one can be a nurse and look after your health, one can work in Harare and bring you money. They are all bringing things into the family, and when I die all those children and grandchildren will come together and bury me.'

Nor shall I forget the wiry Maasai patriarch Simeon Olesaguda, whose 200 cattle and 500 goats graze on the grassy plains and thorn trees of Kenya's Rift Valley. Olesaguda's thirty-five children, by five wives, virtually fill the nearby primary school. Within twenty years, his children and grandchildren will make up a small village.

Not all African men are equally prolific: polygamy is a privilege of age and resources. But Olesaguda's and Chiodza's wives, with an average of seven children each, are entirely typical of sub-Saharan Africa. Though Africa's soils are among the least fertile in the world, her women are the most fertile. The current birth rate – running at 47 births per 1,000 women aged between 15 and 44 – is a full 9 points ahead of South Asia, 14 points in

front of Latin America, and no less than 25 points ahead of East
Asia. Around 1980, the average African woman would have 6.7
children during her lifetime – two more than her Asian or Latin
American counterparts. In Benin, Niger, Senegal, Rwanda and
Tanzania the average is more than 7. In Kenya, most fecund
nation in the world, the average is 8 – four times the fertility rate
in the developed world.

No continent in history has ever seen its population grow so
rapidly. Back in the 1950s, the growth rate of sub-Saharan
Africa was only 2.2 per cent a year – equal to South Asia and
Latin America now. Since then it has accelarated, as death rates
fell but birth rates stayed obstinately high: 2.6 per cent in the
1960s, 3 per cent in the 1970s. The present peak rate of 3.2 per
cent a year is projected to persist until the end of the century.
Latin America's population passed its peak growth rate of 2.8
per cent a year way back in the early 1960s. Asia's peak of 2.5
per cent a year came in the late 1960s.

Before the coming of the Europeans, black Africa's population
was probably around 70–80 million, expanding slowly, but kept
in check by endemic diseases, warfare and (since around 1600)
by slavery. The traumatic early years of European penetration
were probably accompanied by actual declines in population.
But the *pax colonialis* and the availability (albeit limited) of
modern medicine soon brought down the death rate. By 1950
the population of sub-Saharan Africa had reached 171 million,
or just over double its pre-colonial size. By 1978 it had doubled
again. The 1986 population, of around 442 million, is expanding
by 14 million annually, equivalent to a whole extra Ghana or
Mozambique every year.

IS AFRICA OVERPOPULATED?

Rapid population growth is not always a problem. In nineteenth-
century America, from all viewpoints except the Indian, it was
welcome. There is a school of thought, until recently influential
among African governments, which holds that rapid growth is
no problem in Africa, because the continent is actually *underpopu-
lated*. The population density in 1985 was a mere 18.2 persons
per square kilometre, a little more than South America's 15, but

way below Europe's 101 or Asia's 102 people per square kilometre. Surely there is plenty of room for growth?

The argument is an oversimplification, and a dangerous one. The fact is that more than half of Africa is uninhabitable. If we deduct the 13.3 million square kilometres of desert or arid land, and the 2.1 million square kilometres of impenetrable forest, we are left with an area of 14.2 million square kilometres. On this area, *Africa's 1985 density was 39 people per square kilometre – more than Ireland, Jordan or Ecuador.* By 2025, even assuming all the forest were cleared, *Africa's population density on her non-arid area will be level with that of present-day Europe.*

All these figures are *averages* for the whole continent. In fact people are very unevenly distributed, with the relatively drier areas and cooler mountainous zones more thickly populated than the humid zones with their acid soils, their virulent weeds and diseases. At one extreme are countries like Congo, Gabon or the Central African Republic, with 5 people per square kilometre or less. At the other are thirteen countries, with one third of the continent's population, that have more than 50 people per square kilometre. Rwanda is more densely settled than the United Kingdom, Burundi than Switzerland, Nigeria than France.

Africa, it is true, has massive land reserves. The total cultivable land, according to FAO, is 820 million hectares, of which only 220 million, or just over a quarter, was cultivated around 1985. But half of the cultivable land is marginal, capable of giving only a quarter to half of the yields that can be had on good land. A full three quarters of the continent's land reserves are in central and southern Africa. The Sahel and West Africa are already cultivating an area equal to the whole of their suitable land, and their reserves, such as they are, are mainly marginal land.

The most thorough study to date on the capacity of Africa's land resources to feed its populations was published by FAO in 1982. It was based on the best available data on soils, slopes and climates from 117 developing countries, including 51 in Africa. The data were compared with the known growth requirements of fifteen basic food crops, and the crop that grew best in each district was selected. The total yields were converted into calories, and since the daily calorie requirements of humans are known, it was a simple step to calculate how many people could be fed from that production. This potential carrying capacity was then compared with the actual populations. Yields vary according to how much fertilizer, high-yielding seeds, and so on are used, so the study produced results for three levels of input: a low level, using no fertilizer, no improved seeds, and no conservation measures; a high level, with around 180 kilos of fertilizer per hectare, all improved seeds, and full conservation works, corresponding roughly to western European levels; and an intermediate level, halfway between the two, roughly equivalent to farming in Colombia or Sri Lanka.

The continent-wide results of the study at first sight seemed encouraging, and were widely quoted to back the claim that there was no real population problem. With intermediate inputs, Africa as a whole could feed 4,300 million people – equal to the population of the whole world around 1978. With high inputs, the continent could support a massive 12,782 million – more than five times the total of 2,500 million projected for Africa for the year 2100 A.D., when the population should have stabilized. However, Africa's low yields and low fertilizer use place her squarely at the low-input level, and on past trends she will remain there for some time to come. With low inputs, *the entire cultivable area of Africa can support only 1,029 million people – a figure which will be passed by the year 2005.*

In real life, of course, it makes no sense to consider Africa as a whole. Most of the big production potentials are in the humid areas, which do not have, and will never have, the manpower to open up all their potentially cultivable land. This huge theoretical potential will never be realized. It is individual countries that must feed themselves, or fail to. At this level the results give much more cause for concern. Out of 51 African countries, no less than 22, with half the continent's population, had more people in 1975 than they could feed from their own land, even if they used their entire cultivable area.

The study also tried to identify those areas *within* individual countries that were overloaded. Already by 1975 a vast extent of 1,335 million hectares – 47 per cent of the total land area – were carrying more people than they could support with low inputs. The population of these overloaded areas was 184 million – more than twice the number they could actually sustain. All zones with less than 150 days in the year suitable for crop growth were critical, as were virtually all the cooler, mountainous areas. The critical zones (see map, p. 351) stretched right across the Sahel from Mauritania to Ethiopia and Somalia between latitudes 20°N and 12°N, taking in the northern savannah areas of Ghana, Togo and Nigeria. The most densely populated areas of Kenya were critical, along with a crescent along the great lakes from south-east Uganda, through Rwanda and Burundi, to parts of Malawi. A southern critical band of semi-arid zones straddled the continent from the coast of Angola, through Botswana and Lesotho, to southern Mozambique.

These results are alarming. But they are, if anything, an understatement. They assume that all land is used for growing staple foods – no fibres, no vegetables, no cash crops, no livestock feed, no village woodlots. Forests and pastures are allowed only on land too poor to grow crops. The calorie intakes assumed are the basic minimum for health and work, with no allowance for unequal distribution of food. The results are based on climatic data that mostly predate the last dry decade and a half in the Sahel. The study also assumed that all potentially cultivable land – an area four times larger than the present cultivated area – was used to grow food crops.

To make allowances for these factors, we should take the potential production only from *actually cultivated* land, and deduct one third for non-food crops and inequalities. The picture that emerges is far gloomier: an additional 17 countries were critical for low inputs in 1975, bringing the total to 39. Only 12 African countries out of 51 were capable of supporting their populations from their own food production with low inputs. These are almost all in the more humid areas of west and central Africa.

With all these facts in mind, we are better placed to answer the question: is Africa overpopulated? Most of the countries of central and central-southern Africa are clearly not overpopulated, and probably never will be. But most of the Sahel, West Africa and mountainous East Africa are overpopulated in view of the low level of inputs they currently use. Many countries in these areas would remain critical even if they could raise their inputs; for these areas, tackling the growth of populations is just as urgent a task as raising the level of agriculture.

THE PRESSURES FOR FERTILITY

But it will not be an easy job. The uncomfortable fact is that Africa's high birth rates and high population growth rates are not the result of chance or ignorance. It is true that the level of contraceptive use is the lowest of any region in the world. The proportion of women of fertile age currently using any method of contraception – traditional or modern – averages only 11 per cent, half the Arab-world average, against 33 per cent in Asia and 42 per cent in Latin America. But the level in Africa of *knowledge* of any method of contraception is much higher than the level of *use*, averaging 59 per cent of women. The conclusion is inescapable: most African women know at least some methods of contraception, but choose not to use them. They *want* large families. Indeed, in seven out of nine African countries covered by the World Fertility Survey, women's *ideal* family size was even bigger than their *actual* family size. The proportion of women in Africa who want no more children is the lowest in the world, averaging only 13 per cent, and in no country exceeds 30 per cent. Compare this with ranges of 29–74 per cent for Asian countries, and 52–62 per cent for Latin America.

The fact is that Africans plan their families just as actively as everyone else – and they plan large ones. Why? The conventional theory holds that high birth rates are linked to low average incomes, low levels of education, low proportions of people living in cities, and high levels of infant mortality. Certainly Africa performs worse on these indicators, on average, than most other developing regions. But it has higher average incomes and levels of urbanization than South Asia, where birth rates are much lower. And on all factors except average incomes, Africa has seen significant advances. Yet over the past decade, while birth rates fell by 10 per cent in Latin America, 14 per cent in South Asia and a massive 44 per cent in East Asia, there has been virtually no change in the birth rate in Africa. Indeed, in nearly all African countries for which data exist, the total fertility rate – the number of children women have in their lifetime – seems to have *increased*, despite advances in education and child health and increasing urbanization.

Clearly there are special factors at work in Africa. Cultural pressures work to keep the birth rate high. Fertility is highly prized in most African tribes. Numerous children are tradition-ally seen as the most important sign of success and achievement, the proof of a man's virility and of a woman's fertility. Infertility and impotence are widely viewed with fear and horror. These cultural beliefs are adaptations developed over the centuries to the demands of the African environment. Foremost among these is the relative abundance of land: until the last two decades, people knew they could always move on to good land elsewhere if it became too crowded locally, and most people still retain that frontier mentality.

In many countries of the Sahel and East Africa, farmers are already pushing against the limits of cultivable land, but save perhaps in Rwanda, Burundi and Ethiopia, there is still the impression of wide open spaces all round. I made a point of asking many farmers in degraded or overcrowded areas how they thought their children would survive. Some said they could move to this district or that district where spare (though usually marginal) land was available. In Ethiopia the standard answer was, 'The Government will provide more land'; in Kenya, 'There is plenty of land in the national parks'; in Zimbabwe (true in this case), 'There is good land on the white farms'.

Communal landownership – found almost everywhere where land abundance still permits shifting cultivation – is a further factor. Where space still permits, land is allocated according to the size of a man's family. The bigger the family, the larger the area it controls. As it is impossible to build up private wealth in the form of *ownership* of land, the pursuit of wealth, power and status among rural African men takes the form of *control* of the maximum amount of land and its produce. Life is like a huge Monopoly game in which accumulating wives and children is the best way to acquiring movable property and power. Large numbers of children are not simply the symbol of success: they are the principal means of achieving success.

Another environmental pressure leading to high fertility is the shortage of labour in Africa, created initially by the lack of draught animals on farms due to the tsetse fly. Without cattle, human labour is still the only means of land preparation and haulage for most farmers. In a continent where landless labourers are few, children are the main source of additional labour.

The main reason for the recent increases in fertility is probably the gradual demise of traditional or natural checks. Two forms of birth control were widespread in traditional Africa – breast-feeding is prolonged, in some cases up to three years, and delays the return of ovulation; there is also a widespread custom of sexual abstinence, typically for at least a year after the birth of the last child. These two approaches, often in combination, work to space children two or three years apart. Their main aim is the health of the child, but their effect is to reduce fertility below what it would otherwise be. Both practices are on the retreat with the advance of education and urbanization.

Throughout the world, in poor countries with no social security system, children serve as a sort of insurance policy against old age or misadventure: each additional child is an additional unit of insurance. The more insecure the environment, the more insurance people feel they need, and nowhere is that insecurity more pronounced than in Africa. There is the variability of climate and the high risk of crop failure. In the past there was the threat of wild carnivores, and the hazard of tribal warfare intensified by the arrival of firearms and the slave trade. Though these latter risks have receded or disappeared, they have been replaced in many countries by civil and international warfare. Even in times of peace, police and army do not reach effectively into many rural areas, and where they do they often cannot be trusted to keep law and order in an unbiased way. As Australian demographer John Caldwell has remarked, having many children may be the best defence of family interests and security. There is safety in numbers.

The unusual fear of infertility in Africa may well stem from the fact that the risk of infertility – from venereal disease and infections after birth – is very high. Overall, an average of 9 per cent of women may be childless because of infertility, ranging up to 20 per cent or more in parts of central and East Africa. Infertility can strike at any age, so there is a tendency to overcompensate for the risk by having as many children as possible early on.

Finally, the predicament of African women produces a number

of pressures for high fertility. A woman's status, and often her security, depend on her having as many children as she or her husband can support. In polygamous marriages, there may be competition with co-wives as to who can have the most children. As women usually have no secure rights to land of their own if their husbands die or divorce them, their adult sons become the key to their access to land and its produce. Finally, given the massive burden of work in the home and on the farm, women may see additional children as much-needed helpers in the field, or in fetching wood and water.

Governments generally reflect the attitudes of their citizens. Given the widespread desire for large families in Africa, it should not come as a surprise that government backing for family planning programmes has been weaker in Africa than in any other region except the Arab world, with its much higher incomes and resources. In 1983 only 19 out of 43 African governments surveyed thought their population growth rate was too high. 11 governments provided no support of any kind for family planning – 8 of them former French colonies, inheritors of French colonial laws prohibiting family planning. They included Sahelian countries like Chad, Niger and Mauritania. Only 8 governments had, in 1983, implemented measures to reduce fertility.

Even the programmes that do exist are ineffective. A recent survey by demographers Parker Mauldin and Robert Lapham found that out of 32 African countries they surveyed, 30 had very weak or non-existent family planning programmes. In Asia, only 6 out of 22 governments had very weak or no programmes, in Latin America only 4 out of 20. The reasons for the weakness in Africa are partly due to lack of serious commitment on the part of governments and lack of support among the people. The initial emphasis on clinics meant, as in health, that services were inaccessible to most rural women. The recent shift to community-based distribution by locally trusted women is more likely to reach the majority, but here too there are the multiple problems of transport, communications, import barriers, administrative inefficiency and lack of skilled personnel that bedevil almost all development programmes in Africa. An activity that relies on getting dependable and regular supplies of imported goods out to every rural area will inevitably be prone to failure in Africa.

AFRICANIZING FAMILY PLANNING

One mainland African country has made a breakthrough over the last five years. Zimbabwe now has the highest recorded rate of use of contraception in the continent, at 38 per cent of fertile women. And more than 70 per cent of these are using modern methods.

Family planning in newly independent Zimbabwe faced severe problems. Services for whites had been available since 1953, mainly through the voluntary Family Planning Association. In 1966, one year after white Rhodesia's unilateral declaration of independence, they were extended to blacks, but the context tainted family planning with political overtones. The ruling whites inevitably saw the rapid growth of the black population as a threat. The African guerrillas were able to portray family planning services as a white conspiracy to keep the black population down. It was an atmosphere in which old wives' tales about contraceptives – that they caused cancer, or permanent infertility – spread even faster than in the rest of the Third World. Colonial times left a legacy of anti-family planning attitudes.

In 1981, only a year after official independence, a new crisis blew up. The guerrilla leaders, once in government, saw family planning as an important aspect of the health service and wanted a strong say in its management. The Family Planning Association, still run by whites, saw itself as a voluntary body and resisted government interference. In 1981 the government invoked emergency powers and took the association over. Within 15 months, all the top managers had left. Until then, the FPA had relied heavily on the injectable contraceptive, Depo-Provera, but this was banned in Zimbabwe following the controversial link with cancer in trials on beagles. Thus by the beginning of 1983, the association had lost its entire senior management and its principal method.

The turnaround was swift and impressive. The African staff took on the senior management roles. An eminent Ghanaian doctor, Esther Boohene, who happened to be President Mugabe's sister-in-law, became programme director. The programme had full political support at the highest level, and integration with the Ministry of Health began. A substantial USAID grant covered a quarter of the programme costs. The integration with health was particularly important, opening up the whole network of government hospitals and health clinics as potential outlets.

The Africanization of the programme affected not just the staff, but the whole approach. As Esther Boohene explains: 'Family planning is built into the African culture: you have to look out for the positive aspects and build on them. So our initial message was: have as many children as you want, but space them widely like people used to.' The emphasis was put, not on limiting the number of children – that might have put off an African audience – but on using child-spacing as a way of guaranteeing the health of mother and child. Even the association's name was changed to Child-Spacing and Fertility Association, emphasizing African concerns.

The government had no explicit population policy at first, and certainly had no sympathy with the idea of pushing the virtues of smaller families. But actions speak far louder than words, and the government takeover of the service sent out the important message that family planning was something the black government, the people's own government, supported. By 1985 there were signs of an emerging change of attitudes, the first official noises that rapid population growth could be a threat to economic growth, and needed to be braked. The association, significantly, changed its name to the Zimbabwe National Family Planning Council.

By 1985 the programme was one of the best-run not just in Africa, but in the world. It had a strong propaganda arm, putting out radio broadcasts in the two main languages, Shona and Ndebele, holding exhibitions at agriculture shows, sending youth advisors to give talks in schools, colleges and community groups on everything from puberty to forms of contraception or ethics.

But the secret of its success lay in three key factors: accessibility, careful medical control, and rigorous supervision of workers at all levels. By 1985 the association had added to its own 28 clinics a further 122 government and local authority clinics and hospitals. Reaching beyond the clinics out into the rural areas is a network of 520 community-based distributors (CBDs), who give out contraceptives and advice. This wide coverage means that most users either receive their supplies at home, or are within easy reach of somewhere they can get them – the typical travel time to a source, even in rural areas, is only 36 minutes.

Each distributor covers an area with a radius of around 20 kilometres, by bicycle. For every ten or so distributors there is one supervisor with a moped. Each month the distributor gives the supervisor a timetable of her movements. Using this, the supervisor makes unannounced spot checks to make sure the distributors are where they should be, to check that stocks of contraceptives correspond with records of supplies to clients, and to advise on any problems. The system allows little scope for common failings of community distribution programmes, such as bunking off to do farming, and selling or even throwing away supplies to cover discrepancies in the records.

The medical back-up is as critical as the supervision. Distributors are trained to take blood pressure. Women with high readings are referred immediately to the nearest clinic. Other women may be given one month's supply of contraceptives to start with, then they too must visit a clinic for a more thorough check-up to make sure they are prescribed the right type of pill or other method. After that the distributor can supply with enough pills for three months at a time. The medical back-up ensures that side-effects are kept to a minimum, and it gives women confidence. As a result, the drop-out rate among users is remarkably low. Almost half the women taking the pill keep it up for two years. A third carry on for three years.

REACHING THE VILLAGES

Community distribution is the backbone of the programme in the rural areas. The distributors are local people, overwhelmingly women, appointed from a list of three candidates elected at village meetings. They must be at least 25 years old with a minimum of six years' schooling, married with a child and using contraceptives themselves. The job is full-time, with a monthly salary of Zimbabwe $150 a month (US$100) – the national minimum wage.

The programme is impressive on the ground. I spent a day with a distributor in the overcrowded and heavily deforested Kunzwi communal area, about 40 kilometres east of Harare. The distributor here is Netty Chirenda, a pleasant, plump girl in her mid-twenties. We had her timetable, and had no trouble locating her. Her clients act as a further check on her movements: they rely on her for supplies. One woman thought she had been overlooked, and ran after Netty shouting almost desperately: 'You have forgotten me, don't forget me!' The woman, Susan Sendera, had three children – twin boys of six, and another boy of three. Her views were typical of the cautious approach to child-bearing that is spreading fast among younger Zim-babweans, carefully weighing family size plans against family resources. She would like six children eventually, but she has been on the pill for two years. She is worried about the cost of education, so she will wait and see how she manages with three at school before she has any more. Netty asked to see her current pill packet and discovered that Susan had missed a number of days: she had early bleeding, thought it was a period, and jumped to the coloured pills to be taken during menstruation. Netty went through the procedure for 'pill muddlers', and explained from scratch the whole procedure of pill-taking.

Further on, in a compound of a dozen round huts with conical straw caps, I was able to see the gradual shift in attitudes across three generations. The compound belongs to 70-year-old Nyama Dzikiti. He has two wives, Peti, sixty-ish and Muchato, ten years younger. They had ten children each, and all but two survived. Six of the adult children live with them: the eldest daughter, widowed, and her eccentric son, one of the many Zimbabweans traumatized by the guerrilla war; another divorced daughter and her children; and the four eldest sons and their families.

Nyama is now paying the price of his fecundity. He has had to sub-divide his ten acres of land into five two-acre patches, one each for himself and his four married sons. But he is still unashamedly in favour of large families. 'My sons had to pay bride-price for their wives,' he explains. 'It's a lot of money – $150 cash and eight cattle. The person who has paid that much for a wife must have many children, or why should he pay the price? If he has only two or three children, what if two of them die?' He is opposed to modern contraceptive methods – but not to the old traditional approaches. 'We used to plan our families. The method we used most was abstinence, and the next was withdrawal. It was the man who was in control. Our tradition was that, once a child was born, the father wouldn't even sleep in the same house with the mother until that child was old enough to be left alone without its mother. We were breast-feeding up to five years. Then Westerners came here and destroyed our culture. Now the mothers use bottles, the child has lost contact with its mother, and the women control family planning.'

But views are changing all around him. His eldest wife, Peti, is in favour of modern methods. 'What is happening now,' she complains, 'is that if your daughter has a baby, she leaves the little one with you and goes away to work, she doesn't even ask your permission. To avoid that, the women will have to use a modern method.' Three of the six fertile women in the compound are on the pill. One is daughter-in-law Eba Dzikiti. She has only two children and would like six – but she wants to put a space of at least two years before the next child. Her husband works away from home, on a commercial farm, and comes home at weekends. The couple have applied to be resettled: they cannot survive on two acres.

Another pill-user is Peti's eldest daughter, 36-year-old Serin-yeni, who was divorced three years ago. She has eight children, and gets no maintenance from her ex-husband. Four of her children are of school age, but she can't afford uniforms for them. Her eldest boy was in secondary school, but she couldn't pay the fees of $40 a term, and he dropped out. She can't even afford to feed him, so he lives with an uncle. There is no way she could afford to have another child. I asked her how many children she would have had, if she could have chosen. She said three.

Serinyeni's own eldest daughter, 20-year-old Marjory Manda-radzi, was visiting that day. She has one child of nine months. She wants only four children, because she is afraid she wouldn't be able to feed and educate more than that. Her husband works 750 kilometres away, at Victoria Falls, and comes home two or three times a year. They don't use any method of contraception.

'How will you stop having more than four?' Netty asked her.

'I will have the four, then I will take the pill.'

'Do you want babies yearly?'

'No.'

'Would your husband be able to practise abstinence or withdrawal on his trips home?'

'No.'

'Then you will have babies yearly. How can you overcome this problem?'

'The only thing is to take the pill.'

Another convert won.

It is too early to tell whether Zimbabwe's success in spreading modern methods of contraception will translate into lower birth rates. It could be that convenient modern methods might be used to replace inconvenient old methods like withdrawal or abstinence, with no effect on family size. But it does seem likely that the birth rate will begin to come down in Zimbabwe sooner rather than later. When women aged 35–39 were asked how many children they wanted to have, the answers averaged 6.9. Among 15–19-year-olds, the average was 4.6. Most international studies find that the effectiveness of the family planning programme in a country has a strong and direct impact in bringing fertility down.

THE LESSONS

Zimbabwe has a number of advantages over most other African countries. It has the lowest infant mortality and child death rates in sub-Saharan Africa, and the highest rate of enrolment in primary schools for boys and girls. On the other hand, it is less urbanized than the average for Africa. It has a higher proportion of workers still in agriculture than Nigeria, Ghana, Benin or Congo. Its average per capita income – about $740 in 1984 – is pulled up by the very high incomes of whites. The average black income is around $250, well below the African average.

These indicators carry an encouraging message: *African countries do not need to reach high levels of income, industrialization or urbanization before family planning gets going. Improvements in health, especially in infant and child mortality, and in education, especially of women, are far more important – and far more attainable in the medium-term future.*

Zimbabwe's programme has built on African values rather than trying to replace them with as yet unacceptable ideas about small families. Family planning is deeply embedded in African culture, as Nyama Dzikiti reminded us. Most areas still have powerful traditions of prolonged breast-feeding and child-spacing, with the conscious aim of safeguarding the health of mother and child. Family planning has the best chance of success in Africa when it builds on these familiar traditions. Mother-and-child health should be the initial focus: an emphasis on the importance of breast-feeding for a minimum of twenty months, of spacing births by three to four years, and of avoiding high-risk pregnancies before the age of eighteen and after thirty-five. These messages, put across consistently by health workers and politicians, could bring birth rates down and reduce infant mortality at the same time.

Zimbabwe's programme is costly – around $25 per couple per year of protection. This is not excessive by international standards, yet it is still a figure few African countries could afford for nationwide programmes unless all the foreign exchange costs and most of the local costs were paid by aid donors. Costs can be lowered by using voluntary distributors (with some loss of effectiveness) or, better, by making family planning one of the responsibilities of village health workers and birth attendants, providing them with blood-pressure kits and including contraceptive pills among their regular supplies of essential drugs. Even then, problems with imports, and with vehicles and fuel inside each country, may well lead to supply interruptions. Many women will get pregnant and confidence will be shaken.

But it is a mistake to assume that people will not begin to reduce their fertility until they have pills and condoms on tap. Europe's birth rate fell steeply before effective modern contraceptives were widely available. Africans, too, already take their own measures. The 6.7 children that the average woman will have is way below the physiological limit of eleven or more. Africans already limit their families, using abstinence, withdrawal and breast-feeding. These should be complemented with less unpleasant natural methods, especially the observation of cervical mucus, and the rhythm method. These no-cost methods could be taught by village health workers and birth attendants and used as practical material in female literacy classes.

With or without modern contraceptives, Africans will begin to have smaller families either when they are compelled to do so by acute land scarcity – which could be a very long way off in most countries – or when they come to perceive that it is in their interest to do so. This will happen when the insecurity of life is diminished, when the potential impact of drought is reduced by water conservation efforts, improved intercropping and tree planting. It will happen when women's burdens are reduced, and they no longer need to breed children as extra labour power; when wood and water are closer to home and hand grinders are available for grains and roots. It will happen when women are granted equal rights over land, or at least secure rights on widowhood or divorce.

It will happen with the spread of education – of pre-school, primary and secondary education, as children's labour power is taken away and new costs of fees, books or uniforms imposed; and of adult education and literacy, as women's own education improves. It will happen, too, when infant mortality is reduced and the loss of a child becomes sufficiently unusual that parents need no longer over-insure against it.

All these are factors that African governments can influence, using low-cost methods that can be applied even in times of financial stringency. We should not expect miracles: few mainland African countries south of the Sahara will pull off rapid declines in their birth rates of the kind achieved in China, Cuba, or Mauritius. There are too many people around like Samuel Chiodza, Simeon Olesaguda, and Nyama Dzikiti for that. But any progress is better than none, and small reductions in growth rates over the next five or ten years magnify into big differences in eventual total populations.

15 The human resource

Human beings in Africa are caught in a vicious circle of poverty, poor nutrition and ill-health. Disease and hunger co-operate in a destructive synergism. Malnutrition lowers the body's resistance to disease: hungry people fall ill more often. But conversely, many common illnesses in Africa, from gastro-intestinal complaints to parasitic infections, reduce appetite and efficient absorption of food: sick people are more likely to be hungry. Illness and undernourishment lower the productivity of workers and increase the time they take off work. The incidence of both is highest during the growing season, when last year's food stocks are running low and insect vectors and water-borne diseases are thriving in the rains. Precisely when the heaviest effort is required in the fields, crucial days for planting or weeding may be missed due to an attack of malaria, or the time needed to deal with a child's severe diarrhoea. Yields are lowered. Malnutrition and poverty are perpetuated.

The major victims are children. One child in every seven born in Africa will not live to see its fifth birthday. Many of the survivors grow up handicapped, or stunted: a quarter of all children in sub-Saharan Africa weighed less than 80 per cent of the norm for their age in the early 1980s. If malnutrition occurs at certain early stages of brain formation, mental development can be affected. Later in childhood, recurrent episodes of ill-health and undernourishment force the child's body to adapt by withdrawal and reduced activity, cutting down the exploration and interaction with the world by which young minds develop.

The high level of infant and child mortality makes parents have more children than they need, to cover against probable losses. Through this channel, ill-health and malnutrition feed rapid population growth, which threatens the environment and slows progress in every sphere.

Just as in agriculture, African governments attempted to intervene in the vicious circle, but their first steps in health care were patterned on inappropriate and expensive Western models. Spending was focused on hospitals and health centres. Highly trained doctors offered curative medicine based on expensive imported equipment and drugs. The high unit cost limited the services on offer: even in 1980, there was only one doctor per 28,000 people in the low-income countries in Africa – 10,000 more than in other low-income countries. In countries like Ethiopia, Burkina Faso, Chad, Burundi and Malawi, one doctor had to serve more than 40,000 people. These are average figures; in fact the services and the spending are overwhelmingly concentrated in towns and cities, and the rural majority are virtually abandoned. In Ghana in 1980, 90 per cent of the population in rural areas shared 15 per cent of the national health budget. In the same year hospitals absorbed 58 per cent of Tanzania's health spending.

Three quarters of all illness in Africa is thought to be connected in some way with water and sanitation. Yet in 1983 only 34 per cent of Africans had access to clean water, and only 38 per cent to proper sanitation. A mere 26 per cent of rural people had clean water – the lowest of any continent. Though the coverage improved from only 18 per cent in 1970, the total *numbers* without access to clean water rose from 116 million to 175 million in 1983. It is not only the cleanliness of water that matters. The distance of sources adds to women's burdens, especially in the drier areas. In Somalia, women may spend half an hour to one and a half hours travelling in the wet season. In the dry season the round trip averages five hours a day. This limits what can be brought home to drinking and cooking water only. Water for washing hands is a luxury that cannot be afforded.

Life, then, is still nasty and short. Even in 1983, life expectancy at birth in low-income Africa was only 48 years – four years more than in 1970, but a full eleven years less than in other low-income countries. For every 1,000 children born in 1983, 120 would die before their first birthday – half as many again as in

Asia or North Africa, and nearly three times as many as in Latin America.

Even before the drought and the debt crisis, the high-cost approach meant health services could be expanded only slowly. Africa's tightening financial trap of the 1980s made it impossible even to *maintain* existing services. Expenditures were cut, but equipment suffered more than staffing. Drug supplies to health workers dried up. In Ghana, real health spending per person in 1982–3 was one tenth of the level in 1974. Some 40 per cent of the country's water supply systems were out of action for lack of spare parts and fuel for pumps.

DOWN TO BASICS

It has been clear for a decade or more that the conventional, high-cost, Western-modelled approach could not deliver health and other services to the rural majority in Africa within an acceptable time-span, even when finances were not so tight.

Back in the early 1970s, UNICEF and the World Health Organization began to elaborate an alternative approach known as basic services. The concept was developed for all low-income developing countries, but it was particularly appropriate to Africa and has become even more so in present circumstances. The conventional approach asked: given our resources, how many people can we afford to provide with services at the normal high standard? The basic services approach turns the question on its head: given our resources, what standards, what technologies, what level of training must we adapt so we can reach everyone within a decade or so?

The basic services model can be applied in health, water supply, education, housing and other fields. In broad outline the idea is very similar to the approach that Africa needs in agriculture. It uses the lowest-cost reliable technologies available. It disseminates those technologies by way of nationwide networks of community-based workers – elected and often paid by their villages, and given a basic training of anything from one month to a year. Community participation is a crucial element in basic services: it reduces the cost of services by contributing cash, materials, land and labour. More importantly, elected committees or village assemblies choose sites, set priorities, and drum up public support and interest. The village-level worker delivers some services directly, but for the most part acts as a facilitator, stimulating people to help themselves, to become the principal agents of their own development.

In the health field the basic services concept is known as primary health care. This involves the creation of a nationwide service of health agents, one for each village or small group of villages, so that everyone has access to care. The agents are usually parallelled by village midwives – often traditional birth attendants given a basic training in hygienic methods. The midwives may also have functions in health and nutrition education, or contraceptive supply. These specialist workers usually liaise with a village health or development committee which helps to mobilize and educate people and gather resources. Health workers are usually paid a modest salary by government or by the community, or sometimes charge for their services. The emphasis is shifted to prevention: the curative approach – still the focus of most Western medicine – becomes a backstop. To deal with the most common ailments and injuries agents are fitted out with a basic kit of first aid and sterilizing equipment, and essential medicines.

As imported brand-named drugs are expensive in scarce foreign exchange, essential drugs programmes are spreading. These involve selecting a list of up to 200 generic drugs, which are much cheaper, and buying them in bulk, or in some countries manufacturing them locally. Tanzania has the most advanced programme of this kind in Africa. Every month a pre-packed kit of 35 drugs and nine other items goes to every one of the country's 2,800 dispensaries and health centres. Buses, cycles and even mules are used to make sure the supplies get through. The programme has greatly improved coverage – and slashed Tanzania's drug import costs by 50 per cent.

The primary health care approach is not simply a matter of providing care on the cheap for the rural masses, while cities go on swallowing the lion's share of the national health budget. It should involve rebuilding the whole of a nation's health service into a broad-based pyramid where the grass roots get a fair share of the resources. The higher levels of service – the health centres and hospitals – now become sources of specialist back-up, of supervision, training and retraining for the lower levels, and places where more difficult cases can be referred.

REVOLUTIONARY LIFELINES

By the early 1980s, a number of promising technologies for basic health services had emerged. Taken together they promised what UNICEF termed a 'child survival and development revolution', offering rapid but extremely low-cost cuts in infant and child mortality, and improvements in the vigour and mental capacity of growing children and the productivity of adults.

Some of the components in this revolution are virtually cost-free: all they require from government is the network of community-based workers to disseminate them, with the back-up of schools, literacy classes, radio broadcasts and every other channel available. No imports or outside inputs are needed, only information. The people themselves apply the information using materials readily available to them. These elements should form the foundation of any programme, for they are the least vulnerable to import and budget cuts.

Cheapest and simplest of all is the *encouragement of prolonged breast-feeding* for at least twenty months. Not only does breast-feeding save the cost of bottle-feeding: it also confers better resistance to disease and lowers exposure to contaminated water and bottles. Bottle-fed babies are between two and five times more likely to die than breast-fed. Breast-feeding has the additional advantage that it is nature's own contraceptive. Women who breast-feed for 18 months gain a breathing space of 8 to 13 months before they can conceive again. The average breast-feeding period in Senegal, of 23 months, confers an average of 18 months' contraception. The incidence of breast-feeding in Africa is still very high, usually above 90 per cent, but it is not always kept up. Less than half the mothers in Kenya and Ivory Coast, and less than a quarter in Ghana, Niger and Uganda, are still breast-feeding the child at twelve months. Promoting breast-feeding through the media and health workers is not enough: all advertising of infant milk formulas to the public should be banned. Indeed, since bottle-feeding is potentially more dangerous in developing countries than many restricted drugs, there is a strong case for making it available only on prescription, for mothers physically incapable of breast-feeding. Even then, wet-nursing is preferable.

Encouraging child-spacing ranks a close second to breast-feeding for low-cost effectiveness in reducing infant mortality. Studies for the World Fertility Survey found that in north Sudan, children spaced less than two years apart are twice as likely to die before their first birthday than children spaced more than four years apart. In Kenya, over a third of children are spaced less than two years apart, and are twice as likely to die in the first year than the average. It has been estimated that two out of every five infant deaths in Kenya could be avoided if all births were spaced at least two years apart. Child-spacing also improves mothers' health by reducing the burden of repeated close pregnancies. It eases the pressure on women's time, helping to alleviate labour problems during the growing season.

Promoting wider child-spacing and prolonged breast-feeding are by a long way the most promising approaches to improving mother and child health in Africa. They are cost-free or even cash-saving. They are familiar, traditional practices. They promise dramatic reductions in early deaths, and at the same time they help to reduce fertility. No effort should be spared in promoting them by every possible channel.

A number of other promising technologies are virtually cost-free. Properly sited latrine pits; wooden stands for dish-washing; separate huts for young livestock (traditionally kept inside the house); cheap water-filters made of layers of stone, gravel, sand and charcoal in a clay pot – all can make useful contributions. Perhaps the most remarkable new technique is *oral rehydration therapy* to prevent deaths from diarrhoea. The average African rural child may have three to five episodes of diarrhoea a year, during each of which normal growth is halted. Mothers often mistakenly withhold food and even water to reduce messy stools. The child becomes dehydrated and may die as a result. Diarrhoeal episodes may account for more than a third of all deaths under age five – more than a million deaths a year in Africa.

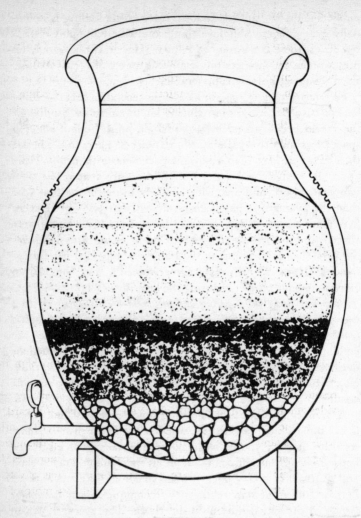

Layers of fine sand, broken charcoal, coarse sand and pebbles in a village pot make an effective water-filter

Oral rehydration was developed in the late 1960s in Bangla-desh: it uses a mixture of salt, sugar and water to keep up the body's fluid levels and prevent dehydration. Studies in Egypt have found that oral rehydration salts reduced pre-school child mortality by 40 per cent. In Turkey the duration of each attack was halved with ORS, and the average child's weight gain per month increased by between 15 and 32 per cent. Oral rehydration salts can be manufactured nationally, or mixed at local health centres. The best approach is to teach mothers to make their own mix from local materials, and to keep feeding their child even when it has diarrhoea. Sugar, salt and water can be used, or the water from cooking certain cereals and vegetables mixed with a little extra salt. ORS programmes exist in a dozen African countries. Probably the most successful to date is in Gambia, where a mass-marketing approach was used to promote home-made rehydration salts. The instructions had to be simple. Mothers were told to use a local soft-drink bottle, Julpearl, available throughout the country. Eight Julpearl bottle-caps of sugar, plus one of salt, mixed in three botttles of water, made a satisfactory formula. Radio Gambia, listened to by three out of every four mothers, broadcast programmes and adverts about ORS, while health workers and nurses toured villages giving demonstrations. Within eight months of the start of the campaign, two thirds of all mothers in Gambia knew how to mix the solution and 40 per cent had started to use it.

The other elements of the child-survival revolution are not cost-free: but their cost is so low and their impact so high that they are well worth the investment – if need be by diverting resources from urban-based curative care. *Growth monitoring* provides mothers with a growth chart for each child – a card on which the child's progress can be recorded after monthly weighings, and compared with curving lines showing the normal progression of growth. Children whose weight increases fall below their normal curve are at risk. Monitoring is a way of drawing attention to them, so mothers can divert more of the family's food supplies to the child, or get medical attention if the child's weight is actually declining. The weighing sessions are better attended the closer they are to the mother's home.

Supplementary feeding helps to improve the growth and resistance to disease of young children. Through vegetable gardening, women (and men) can be taught to grow the leafy green and yellow vegetables that provide important vitamins and minerals, and to combine cereals with protein-rich legumes and vegetable oils to increase the nutritional value of meals. Supplementary feeding can be introduced through community-run pre-school nurseries, as in Bale province, Ethiopia (p. 289). *Vitamin A supplements* can be provided at a modest cost of $0.15–$0.20 per child for a year's supply in capsule form – and can cut deaths of under-fives by almost 30 per cent.

Water supply is one of the single most important elements of environmental health. Provision of clean, accessible drinking water has multiple benefits. The spread of water-borne diseases, from guinea-worm to diarrhoea, is reduced. Washing is easier, so the incidence of diseases like leprosy and trachoma is cut. Washing water can be used to water trees in the family compound. And water points closer to home can reduce women's burdens. Water supply does involve some higher-cost and imported inputs like pumps and cement. But costs can be as low as $20–$30 per pump. Water supply projects that involve the community in choosing pump sites, and above all in managing and maintaining equipment, are far more successful than those with no local participation. One OECD survey in seven African countries found that 40 per cent of water projects involving maintenance by outsiders had breakdowns that lasted over a month – against only 10 per cent of projects where local people were trained to do the maintenance. Clearly, well and pump technology must be selected so that it involves a minimum of outside inputs, and so that maintenance is simple.

Water supply, however, is no easy task outside the humid zone. The water table is often deep and requires deep cement-lined wells for access. After seventeen dry years in the Sahel, the water table has sunk and many old wells need deepening. The burden on women's time is such that unless a new well is closer than one kilometre, women are unlikely to bring home enough water for adequate washing, and in the rainy season, when ponds fill up and streams flow again, they will use a dirty source close to home rather than a clean source further away.

THE COMMANDO APPROACH

Immunization is the last of the low-cost, high-benefit approaches to health. Six major diseases can be prevented by immunization:

diphtheria, tetanus, whooping cough, polio, tuberculosis and measles. Between them, they kill an estimated five million children in developing countries every year and disable an equal number, yet a lifetime's protection costs a mere $5 per child.

Immunization coverage in Africa is better, on average, than in Asia, but still only two out of five African children are immunized against tuberculosis in their first year – and only around one in three for the other five diseases. An effective vaccination programme requires trained staff, working vehicles with sufficient fuel, working fridges and ample supplies of vaccines, syringes, cold boxes and so on. All of these are more problematic in Africa than in any other region. From the mother's point of view, immunization requires a programme of at least three visits, often to a distant site. In a continent where rural transport is often non-existent and women's time overladen with competing demands, it is not surprising if mothers often don't turn out for every session.

Despite these multiple obstacles a number of African countries have made remarkable progress with immunization. Malawi launched a maternal and child health-care programme in 1975 and by 1983 between 60 and 86 per cent of children were immunized against the various diseases – and around half against all six. Whooping cough referrals dropped by one third between 1978 and 1981 and measles cases by two thirds.

One of the most remarkable successes in immunization occurred in Burkina Faso. It is one of the last places you might expect a breakthrough: the fourth-poorest country in Africa, with an average per capita income of only $140. It has the fourth-highest infant mortality rate in Africa; indeed, 19 out of every 100 children born die before their fifth birthday. It has the second-lowest provision of doctors in the world – one for every 48,000 people in 1983. Outside the capital, Ouagadougou, many provinces have only one per 100,000 or more. It is estimated that only one person in four has access to modern medicine.

Late in 1984, Burkina launched a blitz attack on her three major epidemic diseases. Within three months of the decision being taken, three quarters of the children at risk had been vaccinated against measles, meningitis and yellow fever. It was, by any standards, an astonishing achievement. The fact that it was done in such a poor country, and the way in which it was done, contains lessons for the whole continent reaching far beyond the health sphere.

Until 1983, what was then Upper Volta had a succession of urban-oriented regimes which taxed the rural areas and spent almost all their limited budget in the cities. In August of that year, a junior officers' coup led by Captain Thomas Sankara established a radical regime which soon showed that it intended to end the neglect of the farming majority. It abolished taxes on the peasantry, and set out a popular development plan in which participation by the masses was central. Since 1978 the country had subscribed in theory to the idea of bringing low-cost health care to rural areas, but progress had been painfully slow. Immunization against the major childhood killers had to be a key element in primary health care, but every year epidemics of measles, meningitis and yellow fever came along and immunization teams had to be diverted from their broader task. In May 1984 the Ministry of Health launched the idea of a commando-style attack on these epidemic diseases. They wanted to institute it in the summer, when holidaying students could be used, but in July James Grant, UNICEF's executive director, persuaded them to delay. In August, ministry officials visited Colombia, where a similar operation was under way. The Burkina cabinet took the decision to go ahead on 19 September 1984. They set a timetable that at first sight seemed impossible. *Starting virtually from scratch the immunization drive was to begin just over two months later, and to be completed within two weeks.* There was almost universal scepticism, both among donors and local professional staff. Health coverage was sparse, roads were poor, vehicles and medical supplies were lacking, and the people had a low level of awareness of the benefits of immunization. But the government was determined to go ahead, with or without donor support. Most donors remained aloof, but enough came forward to make the operation possible, including China, the Netherlands, and WHO. The major donor was UNICEF, which paid 28 per cent of the costs, while the Burkina government paid about half.

The task of organization was formidable. A national co-ordinating committee was formed, with representatives of most major ministries (including defence, education and agriculture), of donors and non-governmental organizations. Regional and local committees were set up. A census of needs and resources had to be carried out: how many children were involved; how many vaccines were needed; how many trained people, injectors, vehicles and other resources were available; how many more were required; what distances had to be covered and how much petrol was needed. The vaccines are sensitive to heat, so 'cold chains' had to be planned, with freezers, fridges and cold boxes, to make sure that vaccines were kept cool and effective. It was decided to restrict the diseases covered to three – measles, meningitis and yellow fever. All three could be injected with high-pressure airjets called pedojets. The training involved was simpler than for syringes, and there were fewer risks of complications and side-effects which might have reduced people's confidence in vaccination. All three could be covered with only one shot each.

Personnel had to be trained. Burkina used a pyramidal system which greatly speeds up training and reduces its cost. Regional trainers are trained in the capital, then return to regional centres where they train lower-level staff. Touring timetables were drawn up. The most crucial task of all was the mobilization of the people. Without that, all the rest of the effort would be wasted. There were radio programmes and posters in local languages; there were travelling theatre groups. But in a country with poor roads, where only a minority have radios, person-to-person communication was central. The key agencies in creating awareness and mobilizing people were the Committees for the Defence of the Revolution. The CDRs were set up in 1983 to protect the revolution against counter-attack, but they soon acquired much broader functions. They have elected officers, and membership is open to anyone. The committees carry out government directives, organize mass labour parties, and act as channels of education and consciousness-raising on all issues from environment and agriculture to health. The CDRs and their affiliated women's groups spread the message about the value of vaccination and organized popular fund-raising events. They carried out the census of children, gave out vaccination cards, and made sure that people turned up at the right place and the right time for vaccination.

The response was overwhelming. Mothers almost took the

vaccination points by assault. They walked long distances, and formed queues often more than a kilometre long, waiting whole days and nights for their turn. In many areas the demand far exceeded the carefully calculated supply of vaccines. The propaganda was so effective that children outside the target age groups turned up, and in border provinces families from Mali, Nigeria, Togo, Ghana and Ivory Coast flocked in.

The impact was profound. In immediate practical terms around 2 million children – almost 30 per cent of the national population – received at least one vaccination each. Previously only 11–19 per cent of children had been immunized against the three diseases: after the campaign, the coverage rates shot up to 60–70 per cent, and the rural coverage was almost as high as the urban. The usual annual epidemics of measles and meningitis did not take place in 1985: somewhere between 18,000 and 50,000 child deaths were prevented.

More important still were the long-term effects. The whole population was made more aware of health issues, and of the fact that they themselves could do something about health. A demand was created for further vaccinations and for health care in general. The commando campaign served as the foundation for an extended programme of immunization, and offered a model for the creation of a comprehensive network of primary health care.

The momentum was still rolling when I visited Burkina in November 1985, a year after the commando. I went to a vaccination session at the village of Rouko, in Bam province. To get there we had to negotiate a rough dirt highway, then strike off into the bush where the road marked on the map was no more than a footpath. We had only a rough idea of the whereabouts of the immunization team, but could hardly mistake it when we arrived at Rouko. Under the shade of the mango trees by the local school, perhaps 300 women were patiently queueing with their children. Some of them had walked from as far as 25 kilometres away. The local Committee for the Defence of the Revolution had been informed of the time and place, and sent messengers on foot or bicycle to surrounding villages. CDR volunteers also handled the business of registering children and stamping their cards, so that the medical team's valuable time and manpower were reserved for immunization.

The Rouko CDR is also building a house for the schoolteacher, a literacy centre, and a new health post. It has made a brave attempt to plant trees on the bare and barren land of the village – but the forestry department supplied the seedlings in August, at the end of the rainy season. By January, Rouko's wells have dried up and the women have to walk 10 kilometres to get water. There is none to spare for watering trees, so most of the seedlings were dead. The CDR's biggest project is to build a dam for drinking-water and irrigation.

ONE VILLAGE, ONE HEALTH POST

The rest of the conventional health system was in a bad way, with poor coverage, lacking the most basic of materials. There is only one fully qualified doctor in the whole province of Bam, to cover a population of 180,000. The largest medical centre, in the provincial capital, Koungoussi, has no facilities for surgery or blood transfusion. Difficult cases have to be taken in the province's single ambulance, along 110 kilometres of dirt road to Ouagadougou. In practice most cases, even serious ones, get no attention at all.

Rouko village has one of only seven health centres in the province, staffed by two nurses. The delivery table in the two-room maternity unit is a raised cement block with iron handles. There are no beds – two women with new-born babies were sleeping on mats on the floor. The pharmacy had only a handful of drugs, for malaria and headaches, and antibiotics. The small petrol-powered fridge was out of action for lack of fuel. Forty or fifty dirty syringe needles were waiting in a pan to be sterilized: again, there was no fuel for the burner.

The vaccination commando has now inspired a wider primary health care campaign in Burkina to plug the gaping holes in the health service and to extend the coverage to 85 per cent of the population. At the base will be 7,500 village health posts – one for every 1,000 people, each one staffed by a village health agent and a traditional midwife trained in modern hygiene. The health agent is chosen by a village assembly. After a month's training course, he or she will be responsible for health education, promoting sanitation and clean water, and first aid, and will be kitted with a few basic medicines. Each village will have a health committee which will help with the management, upkeep and finance of the health post. It is up to the village to decide how health workers are rewarded. In some cases they get a stipend of $50 a year, in others they keep the cash from selling medicines, or are given a couple of sacks of millet for their trouble. The birth attendants will also be trained in basic family planning, and provided with barrier contraceptives.

At the second level there will be 450 Centres for Health and Social Promotion with two nurses, one mobile health worker and a midwife. These centres will be responsible for supervising the village health workers, and for immunization. Above this will come 59 medical centres, and at the top level 10 regional hospitals. The higher levels will have the roles of supplying, supervising and training the lower levels, and of dealing with more difficult cases referred from below.

The timetable for the 'one village, one health post' campaign was just as ambitious as the vaccination commando. The formal decision was taken in August 1985, with the aim that the whole system should be in place within five months. Again there was a formidable checklist to get through, starting with the organization of village health committees and the election of agents, through the equipping and building of health posts, to the training of trainers who would train the agents. During my visit (November 1985) I passed dozens of villagers out on Sunday, digging clay, making bricks, and building modest health posts. Village health agents were two weeks into their course which would be followed immediately by two-month courses for traditional birth attendants.

The primary health care campaign is even more ambitious than the vaccination commando, and there will, inevitably, be more problems. The difficulties with supplies of drugs, fuel and spare parts for vehicles, fridges and sterilizers, will persist for as long as the country's foreign exchange position is so weak, or until donors agree to shoulder on-going costs in a way most have shied away from. The training courses are painfully brief – for the health agents, a mere fifteen days of basic instruction, followed by eleven days of practical work and revision. Many of the health agents are illiterate, and keep their records by visual symbols – a large circle for a wound, a small circle for a malaria tablet. There will be problems of false diagnosis, of wrong treatments, of inadequate supervision and poor records. But this venture is a long-term one. Close supervision and in-service training will be essential in maintaining and up-grading staff standards. The crucial benefit of the campaign is that it will put into place once and for all, and all over the country, the basic health-care structure that had so far been delayed, and would have gone on being delayed for decades at the normal pace of things. The existence of the structure will shift the whole balance of government health spending away from the apex towards the base of the pyramid, away from privileged urban dwellers to the rural majority. It will create a popular demand for health services which will lead to pressures for improvement. And it will create an interest in health actions that people can undertake for themselves, at very low cost or no cost, from digging latrines and refuse pits to home-made water-filters and rehydration mixes.

The vaccination commando was a seminal experience for Burkina Faso and could serve as such for the whole of Africa. It was an exercise in confidence-building through action. It showed everyone in the country, from the Cabinet down to the remotest village, just what could be achieved if everyone pulled together. Political commitment by government was an essential ingredient. The timetables were perhaps a little too impatient: a target of one year might have been wiser for primary health care, to allow more time to train agents and to secure long-term supplies. But a tight schedule, provided everyone is serious about it, has one massive advantage over the usual approach of gradual and slow progress. It creates an inescapable obligation, which cannot be fudged or indefinitely put off.

Manpower shortages are always a handicap, but the pyramid approach to training – beginning with the training of trainers, and so on through provincial and district level, right down to village level if need be – allows a very rapid solution. A nationwide network of village-level workers can be put in place within three months to a year, depending on the length of the basic course. This approach can be used in health and family planning, in literacy, in pre-school education, in conservation and agricultural extension.

The other, equally essential ingredient in Burkina's vaccination success was community mobilization and mass participation, guided by a nationwide structure of popular organizations. The lesson is clear and immensely heartening: when real political commitment and mass participation combine in well-directed actions, almost anything is possible in Africa.

A HEALTH AND POPULATION REVOLUTION

Pre-requisites of all stages

1. *Nationwide dissemination network*
   Community-based health
     agents
   Traditional birth attendants

2. *Use of all channels of
     communication*
   Schools, literacy classes
   Media
   Political organizations
   Community groups, etc.

3. *Community participation*
   Individual self-help
   Village committees

Stage One

*Technologies requiring only a nationwide dissemination network, no cost to villages
other than labour*

To improve health and reduce
fertility:

- Encouragement of breast-
  feeding (20 months +)
- Child-spacing (2–3 years +)

by means of natural methods:

- Abstinence
- Cervical mucus observation
  and rhythm (where female
  literacy adequate)

To improve sanitation:

- Pit latrines
- Wash stands
- Separate quarters for
  livestock

To improve nutrition and health:

- Vegetable gardening
- Solar drying of vegetables
- Nutrition education
  – green leafy vegetables
  – combining cereals, legumes
    and oils
- Home-made oral rehydration
  salts
- Supplementary feeding (local
  produce) in pre-schools

Stage Two

Technologies requiring low-cost external inputs

- Charcoal, stone and sand water-filters (tap only)
- Vitamin A supplements
- Growth monitoring by child-weighing groups

Stage Three

Technologies requiring higher cost and imports plus more sophisticated management

- Clean water supply
- Immunization
- Essential drugs
- Contraceptive pills injectables and condoms

16   By their own bootstraps: mobilizing for change

So far we have been dealing with projects aimed largely at particular sectors – agriculture, conservation, forestry, energy, population, health. We have seen all along just how closely all these sectors are interwoven both in Africa's problems and in the potential solutions. We have also seen how important popular participation is in every sphere.

If it were possible to use, or to create, an organ of popular participation that could work on a broad front across all areas, the benefits would be powerfully multiplied. The feeling of confidence, of control over one's own destiny, gained in one area could be applied in others. The openness to change created by one kind of successful venture could be exploited for others. The ability to mobilize people, developed for one activity, could be used for many more.

Of course, the traditional African village *was* such an organ of popular participation, an allocator of land, an organizer of community labour. But the colonial state removed the effective powers of the traditional authorities, and the successor, independent states for the most part, did not restore them.

Africa is weaker than any other continent in popular grassroots organizations of every kind, from trade unions and pressure groups to voluntary organizations and societies with development purposes. All too often, the state monopolizes the activity of development while popular organizations are no more than social clubs, secret societies or religious sects. Grassroots organizations tend to grow with the general level of income and especially education, since literacy and familiarity with laws and procedures are needed when dealing with the modern state. As landholding in Africa is still relatively equal, social divisions are less pronounced, so organizations based on class are weaker. Ethnic divisions cut across and weaken groupings which share a common interest.

There are exceptions. A few countries, like Kenya, have a plethora of grassroots self-help groups. In Kenya's case the groups often start as savings clubs, helping members to buy tin roofs or pay school fees, and then move on to higher things, building nursery schools, starting tree nurseries, or co-operating to terrace each other's land (see pp. 121f and 202). In Kenya's case the self-help or *harambee* movement had its origins in the guerrilla rebellion against British rule, as a way for women to overcome the problems of having menfolk away at war. Women remain the backbone of the self-help movement in Kenya. In Zimbabwe, farmers' groups, who share labour and tools (p. 92f), could become all-purpose voluntary groups of a similar kind, as the members of the Organization of Rural Associations for Progress, ORAP, in the south of the country, already have done.

In other countries the grassroots grouping is statutory or state-backed. Burkina Faso's Committees for the Defence of the Revolution (p. 269), which mobilized people for the vaccination campaign, are a good example. Almost everywhere the CDRs have become genuine mass organizations undertaking a broad range of development activities, from tree-planting and the control of tree-cutting and grazing, to road repair and the construction of village health-posts. At present there is a little too much emphasis on carrying out directives from above; it has worked so far because the government commands general goodwill after abolishing the rural poll-tax, and so far most of the directives have been well chosen and widely seen to be necessary and beneficial.

DEVELOPING WITHOUT DESTROYING: THE NAAM MOVEMENT

One of the most successful of the voluntary groupings is also in Burkina Faso: the Naam movement, centred on the Yatenga plateau in the north of the country.

The founder of the Naam movement, Bernard Lédéa Ouedra-ogo, was a teacher who moved into rural development work in

the early 1960s. His task then was *animation* – a combination of
education, consciousness-raising, motivation and organization of
peasants to undertake development actions. He remembers the
difficulties of that approach. 'I would go into a village,' he told
me, 'call a meeting and tell people we needed to organize a co-
operative. But co-operatives were a French concept. It was only
a few years after independence. They still regarded any French
idea as a form of colonial domination.'

 Ouedraogo, frustrated by his lack of progress, started to think
of alternative approaches. The aim remained development – but
indigenous development, fuelled from within, not pulled from
without. It had to be an improvement on traditional values and
methods, but it should build on the best of these, not set out to
undermine and destroy them, as Westernizing approaches did.
With these principles in mind, Ouedraogo developed his fundamental philosophy of development: 'To make the village responsible for its own development, developing without destroying,
starting from the peasant: what he is, what he knows, what he
knows how to do, how he lives, and what he wants.'

The logical consequences of these principles led Ouedraogo to
many of the secrets of success shared by our other projects.
Because the peasant is poor and has no access to imported
equipment, Naam's activities are wherever possible low-cost,
relying on local tools and materials. They use low-cost purchased
or imported inputs only where there is no alternative. Respecting
the peasant's knowledge and experience means building on
traditional concepts and technologies, rather than trying to
supplant them at one blow by alien and unfamiliar Western
approaches. Starting from what the peasant wants, and making
the village responsible for its own development, means maximum
popular participation in the aims and means, in the management
and maintenance of projects. Indeed, the Naam movement has
taken participation to its ultimate conclusion: *all the activities it
has stimulated are managed and run by villagers. Professionals and experts
take up their proper role of training, technical and financial back-up.*

Ouedraogo's basic method gives the Naam movement an
astonishing talent for backing winners – ultra-low-cost technologies, using local materials, that can be taught and spread quickly.
The groups have taken up Oxfam's water-conserving technique

of building stone lines along the contours, and are spreading it as fast as Oxfam itself. They have adopted the improved three-stone stove and have been pushing it longer than the government. Their model of the elected village health agent backed by a village health committee has been seminal in the government's plans for primary health care.[1]

Ouedraogo is a teacher by origin, and it is his pedagogical techniques that help to explain the speed by which successful solutions spread. The animators are not outsiders, government employees reluctantly serving their time in the 'bush' like a prison sentence. They are elected by the people, and share the same circumstances and culture. Techniques are spread by the channel African peasants trust best, the proof of their own eyes. Visits are arranged to take people to villages where new methods are in successful use. And much practical training is undertaken in 'workplace schools' (*chantier-écoles*) where people learn by doing.

The Naam movement started in 1967. The name derives from the traditional Mossi village organization which Ouedraogo chose as his starting-point, a grouping of young men and women which was formed each rainy season to help with planting and harvesting. It was a traditional self-help co-operative movement, based on equality, and with elected leaders, which villagers were familiar with. The chief offices were Kombi Naaba – leader and spokesman; Togo-Naaba – herald and executor of orders; and Rasam-Naaba – guardian of the treasures of the Mossi emperor. Ouedraogo took the traditional titles for the President, Secretary and Treasurer of the new associations. In each village the elders' agreement to these changes was secured. By 1985 there were no less than 1350 Naam groups, grouped in seven regional federations. Similar organizations, based on traditional self-help institutions, had been created in Senegal, Mauritania, Mali, Niger and Togo.

In 1976, Bernard Lédéa Ouedraogo set up an umbrella organization, the Six 'S' Association, to provide technical and financial help for the Naam groups and to raise international funds for indispensable imports such as modern medicines,

[1] See pp. 165-70, 217–19 and 271–3.

cement, pump equipment and improved hand tools. Six 'S' stands for *Se Servir de la Saison Sèche en Savanne et au Sahel* – making use of the dry season in the Savannah and the Sahel. The long dry season, six to nine months when there is not enough rain for crops, is one of the greatest problems of the semi-arid areas – yet, paradoxically, one of the greatest resources, a massive reservoir of under-used labour. In the good old days these months were used for festivities; in more recent years they have been used for migrant labour with many adult men heading for the Ivory Coast. But they represent a vast potential for conservation and development work, and this is how the Naam movement uses the dry season.

The Six 'S' headquarters in Ouahigouya, the capital of Yatenga province, is a constant bustle of activity. Lorries roar in and out with deliveries of pick-axes and shovels, wheelbarrows and watering cans. Under the shade of straw pergolas, a dozen workers twist heavy-duty wire into meshes to be used for fencing or for gabions – cages of rocks used in dams. Beside the regional federation's warehouse, another team of three make fuel-saving aluminium cooking pots in traditional shapes. Battered old kettles, blackened pans, bike pedals and sardine cans are melted down on a charcoal furnace, and poured into a mould of wet sand and clay.

But the heart of the Naam movement is where it should be – in the villages. Naam groups have built wells and dams, set up vegetable gardens, planted village woodlots, built village shops and mills. Most villages have embarked on a number of these ventures – many on a virtual revolution of self-help development across a broad front. The Naam group in the village of Ouffré only started up in 1979, but in its first six years villagers dug a well, saved up to buy a millet mill to save the long trek to the nearest town, and a communal donkey cart to carry their produce to market. They planted a two-hectare woodlot and started rearing chickens. Virtually every household has made itself one or two improved three-stone fires (p. 217) to save fuelwood, and simple water-filters from layers of charcoal, sand and gravel in a clay pot. They have a village health agent, who earns her fees by selling basic medicines for malaria, worms, eye troubles and headaches. The Naam movement has introduced modern medicine without attempting to stamp out traditional healing; indeed, it fosters village pharmacopeias. In Ouffré, Kadisso Ouedraogo combines both roles. Before being elected as the village health agent she had been a traditional herbalist for 46 years, and keeps a treasured collection of dried flowers, seeds, roots and barks. Some of these are probably effective; some work, if at all, by the placebo effect. For malaria she recommends a herbal decoction; for itching sores a lotion made from a herb fast disappearing due to desertification; for gonorrhoea, the ground ashes of a bush rat; for infant fevers a powdered tree-root mixed with charcoal dug from a place where no corpse has passed.

The Naam group in the village of Somiaga has another village pharmacy, a mill, wells for drinking-water, a tree nursery and a woodlot. In 1983 Somiaga started a cereal bank. Previously, villagers had to buy grain from the market towards the end of the dry season, or during the first two months of the rains, when their own granaries were empty. Cereal prices were highest then, and often people went into debt at usurious rates. At harvest time, when grain prices were lowest, they had to sell their grain to pay off the debts. The cereal bank buys cereal in cheaply at harvest time, from inside and outside the village, stores it locally, and sells it at cost when it is needed. Cereal banks cut out the middle man's profit, provide more favourable prices for growers and consumers, even out the seasonal switchback in prices, and ensure that grain stores are available locally. They are an important source of food security.

When I was there the people of Somiaga and neighbouring villages were putting the finishing touches to what is the pride of the Naam movement – an enormous dam, 180 metres long and 4 metres deep, that will eventually irrigate 50–60 hectares of rice, potatoes, vegetables and fruit trees. It will raise the level of water in local wells, make it easier to water tree seedlings in their crucial first year, and provide an ample source of fish.

The Naam animator for Somiaga, Madi Ouedraogo, told me that in the recent run of dry years ten whole families had left the village, carrying all their possessions on their heads, heading for the wetter south-west provinces. 'We saw that in the dry season our people had nothing to drink, our animals had nothing to drink, and our wells dried up. But in the wet season we watched the water rushing by, sheeting over the dead land, roaring down the gullies, washing away.' In 1981 they started to build a dam without any outside help, from piled-up stones. But a heavy storm in the rains of that year washed it away.

They asked Six 'S' for help. The association brought in an engineer. An earthmover was hired for the two wings of the dam, but the central stretch of 65 metres was built by hand, save for the concrete spillway. It is a vast structure made of 2,300 large rock-filled gabions stacked eight deep. Villagers came from Somiaga and the sister villages of Becka, Laona, Sissemba, San and Bouri. From November 1982, work proceeded every dry season, three days a week. On the best days up to a thousand people came out – old women loading baskets in the gravel pits, or preparing the communal meals; men, women and even children carrying stones on their heads for eight hours a day under the baking sun. Up-stream, where floods gathered off the barren slopes into rushing rivers, check dams were built in the gullies to slow down the flow and collect soil, so the dam would not silt up quickly. 'Allah smiled on our labours,' Madi Ouedraogo proclaimed. 'There was not one scorpion bite, not one snake bite, not one pick-axe wound. Allah wanted us to build this dam.'

The dam now stands as a source of pride for everyone who laboured on it, incontrovertible proof of their ability to control their circumstances instead of being controlled by them. The women called it 'throwing stones at the drought'. A more rhetorical visiting government minister said it was 'an astounding example of the people's determination to conquer imperialism and underdevelopment'. Minata, a Somiaga matriarch and veteran of another dam at Saye, composed a song: 'All the men who travelled to Mecca say they gathered stones to throw at the evil tombs of the disbelievers. Like them, we gathered stones. But we are going to build a dam, a future for our children, our village, for Burkina Faso and for all of Africa.'

RIDING A WAVE OF PROGRESS

Voluntary groups have a clear edge over official groups when it comes to popular mobilization: members belong because they actively want to belong, and the group's activities are not in any way dictated from above. But official groups can form the engine of development across a broad front, as long as they are genuine community organizations that command popular confidence.

Ethiopia's Peasant Associations are a prime example. Before the 1974 revolution, Ethiopia had the most oppressive and unequal system of land ownership in Africa outside South Africa and Rhodesia. Most peasants were sharecroppers, paying half to three quarters of their produce to the landlord, providing sheep for the landlord's feasts, or contributions when his taxes were due. The peasant's wife was the unpaid servant of the lord's wife, his children the lord's unpaid shepherds. The feudal lord was also the judge in his own cause.

The emperor Haile Selassie's inactivity during the famine of 1973–4 sparked off the revolution that put an end to feudalism. Countries that have actively liberated themselves from an internal oppressor or a colonial ruler have a unique opportunity. The energy and activism which is released by struggle and liberation has a forward momentum that can be carried into the broad process of development. This momentum explains the strength of self-help in Kenya, and the extraordinary dynamism of villages in Zimbabwe or Burkina Faso. It is also a powerful factor in Ethiopia.

Two early reforms of the Ethiopian revolution created the pre-conditions for a broad movement of self-improvement. The first was a radical redistribution of property rights. Every peasant family was given the use of one to five hectares of land, and feudal dues and shares were abolished. In the centre and south of the country, where feudalism was strongest, peasants felt for the first time that they were in a position to control their destiny, rather than being in the hands of forces beyond their power.

Ownership of land is now vested in the state, but control and allocation lies in the hands of the Peasant Associations to which all farmers belong. The associations elect their leaders, and do much of their important business by way of general assemblies of the whole village. They represent members' wishes to local government, and also act as the lowest, unpaid tier of government, organizing work parties, administering basic justice, and so on. In the north of Ethiopia, where feudalism was less pronounced, there is some resentment of the collective ownership of land and some resistance to the associations. But in the centre and south of Ethiopia, the Peasant Associations are genuine community organizations, commanding popular support, with significant local powers and an all-purpose brief. They provided the framework by which the feeling of activism created by the revolution could be harnessed for community goals.

The UNICEF-backed Regional Integrated Basic Services (RIBS) project in Bale province in southern Ethiopia made full use of these advantages to pioneer a new participatory relationship between popular organizations, government, and aid donors. The project had its origins in the most troubled time in Bale's recent history. In 1977, while Ethiopia was torn by internecine political struggles, Somalia invaded the Ogaden in the south of Bale. The following year, drought hit the region. The two disasters created a massive problem of refugees and destitutes, but also created an opening for radical change. To cope with the crisis, the Relief and Rehabilitation Commission took on the task of co-ordinating all agencies, resources and government ministries in Bale. Hundreds of new villages were built, and other dwellings that had been scattered and isolated were gathered into compact settlements for better defence. This move made it easier to provide services like schools, health posts and water. The co-operation between ministries in reconstruction provided an opportunity to develop government services in an integrated way. UNICEF, which was already involved in emergency assistance, spotted the opening, and the Bale Integrated Basic Services Project started in 1980.

The project is based on participation and response to popular demands. The focus is the Peasants' Association of each village and its development committee, which also includes representatives of women and youth. The project's grassroots workers, known as co-ordinators, encourage villages to examine their problems and needs, and to decide on priorities. Villages are expected to contribute as much as they can themselves, in labour, land, materials or cash. If they need outside assistance, they submit their plans to the local RIBS committee, made up of representatives of local Peasants' Associations, government departments, and RIBS project staff. The committee looks at all the local requests, and decides which ones it has the staff and resources to accede to that year. The proposals then work their way up a hierarchy of steering committees at district, regional and national level, to see what can be afforded. UNICEF is then approached for funds and materials to bridge some of the gaps between local demands and existing resources.

The usual relationship between government and citizens is reversed. Government departments no longer plan the extension of services as they see fit, without reference to other departments. The steering committees co-ordinate all development work – so, for example, water supplies for drinking and for irrigation are planned together; schools and nurseries incorporate health and nutrition measures, and so on. The emphasis is on what people can do for themselves. The role of government is to provide technical and limited material support. The role of the donor, UNICEF, is to supplement Ethiopian resources to permit developments that might otherwise have to wait for years. Government and donor no longer dictate what is done, but facilitate what the people want done. The outcome has been, in UNICEF's own words, 'the creation of an integrated set of essential services that are geared to the genuine needs of communities and are low-cost, self-sustaining and flexible to changing needs.'

The project has provided very good value for money. Aid has acted as a catalyst, bringing out a much greater quantity of matching local effort, for an annual outlay of only $1m a year. Conventional services cannot hope to reach the majority of people in any foreseeable time span. There is only one hospital for the whole region's 128,000 square kilometres of the region, and only nine doctors (four of them Cuban) for its one million people. The RIBS project has vastly extended the coverage by using low-cost technologies and low-paid or unpaid workers. No fewer than 573 community health assistants have been trained and kitted out with a set of basic medicines and first aid materials. The agents are volunteers, elected by their communities and trained for three months. They are unpaid, but neighbours help out on their fields at crucial times of the year. Perhaps their most important function is as health educators: they teach mothers about oral rehydration for diarrhoea (p. 263), and persuade people to dig latrines and refuse pits. Three out of five people in the region now use pit latrines. Some 328 traditional birth attendants have been trained and kitted so they can deliver babies with far greater attention to hygiene. Clean water is a key sector in preventing disease and malnutrition: over 270 springs have been cleaned and protected, and 37 protected wells dug. Nutrition has also been improved by a new focus on irrigated gardening, growing dark-green leafy vegetables and carrots to supplement the staple cereals; 25 rivers and springs have been diverted to irrigate 530 hectares of gardens.

Education has made great strides since the revolution – there are now 314 primary schools, more than four times the pre-revolution total, and enrolment is up from 10 per cent of the age group to 42 per cent. But that still leaves almost half the villages with no school, and three out of five children with no schooling. Here too the basic services approach is filling the gaps. Where there is no school, children can attend one of the region's 239 reading centres, or its 211 listening clubs following radio courses in local languages.

On the ground most villages are involved in several of these areas simultaneously. Indeed, many have embarked on a total process of development that has brought them very far from the serfdom of pre-revolution days, or the destitution of 1978–9 after war and drought.

The village of Hora Boka sits at the foot of the Bale mountains, three kilometres behind the town of Robe. The houses and school were built in 1977, when people came together for defence against the Somalis. That year they lost all their cattle to raiders. In the drought that followed they were reduced to eating wild fruits from trees on the hills. In 1979, after their first decent harvest, they built the village assembly-hall, a vast cavernous building walled with bamboo matting. On the day of my visit, the farmers were meeting to decide how to organize the harvest and collect taxes. The same year they elected a community health agent, Siraj Hassan, and built him a small health post where he sees patients. He opens three hours a week for consulting, and once a week supervises mothers as they weigh their infants and fill in their growth charts (p. 265). He has persuaded 500 of Hora Boka's 585 households to build pit latrines – the rest are waiting for the timber to complete them.

1980 was a busy year. The people built offices for the Peasants' Association, the Women's and Youth Associations, and the people's court. They started up a service co-operative, which sells farming inputs and basic goods, and built a shop for it. They constructed a reading room – a big, mud-floored hall with low wooden benches, stacked at the back with racks of simple reading material on practical topics like preventive health care, nutrition and improved farming. The literacy teacher, 33-year-old Selan Aliyi, has six years of schooling. He was wounded in the Somali war, and limps badly.

In 1981 they built a kindergarten, with $1,500 from UNICEF for a tin roof and cement. The kindergarten takes children from age four and five. The teacher, elected and paid for by the community, uses simple locally produced equipment to get children used to working with shapes, sounds, letters and numbers. The Peasants' Association set aside 20 hectares of land for the school. Some of the produce is sold to pay the teacher's salary, some of it is used to give the children a nutritious midday meal. The local school says the kindergarten graduates are infant prodigies: by the end of their first year, most of them have been promoted to grade three. Also in 1981, the people of Hora Boka invested in a flour mill which charges 40 per cent less than the mills in Robe and saves women the six-kilometre round trip. And they started planting trees. Their woodlot now covers 20 hectares, and they are planting trees round all houses and communal buildings.

1982 was women's year. The villagers dug five wells, to reduce the chores of water-gathering. And the women's association started up a whole range of activities designed to provide a cash income for women, or to reduce their expenses. Four women work 25 hours a week knitting and sewing, on machines paid for by UNICEF. The garments they make are sold in the co-op shop, and proceeds are split 50-50 between the workers and the women's association. Six more women make village soap from a mixture of caustic soda, animal fat and borax. The bars are sold through a network of village co-op shops, and are in increasing demand.

In 1983 the villagers embarked on improving their houses. The traditional house is a smallish circular mud hut, with a single, windowless room into which are crammed beds, sacks of grain, pots and pans, an open fire, and young animals. The improved house is much larger, with a high roof of the kind once reserved for feudal lords. Inside, it is partitioned into four: a lounge, with mud benches covered with hides, a store-room, a bedroom, and a kitchen. There are windows to let air and light in and smoke out. Animals are kept outside, in a walled shelter under the lee of the roof. So far only 40 improved houses have been built – it is a slow process, because of the shortage and high price of wood for construction.

Looking at Hora Boka's neat rows of houses, the impressive civic centre with its offices, mill, workshops, court, assembly hall, health post and kindergarten, it is hard to imagine that only eight years ago its inhabitants had been reduced to total destitution. I asked the secretary of the development committee, Getachew Zeleke, how he explained the extraordinary dynamism of the village. 'In the past,' he answered, 'we were scattered, and nothing was according to our will. Now we are together, and there is no barrier. We must be strong and lead ourselves.'

FIGHTING AGAINST THE ODDS

Not all villages have made such smooth progress. Below the Bale mountains, beyond Robe, the land slopes more gently. Alesedestu lies on this slope, overlooking the village lands on a broad and almost treeless plain. The village's focal problem is water. The soil is sandy and there are no impermeable layers near the surface that could hold sub-surface water. In the rainy season villagers obtained water from the same insalubrious, muddy ponds as their cattle. In the dry season they had to walk ten kilometres or more to the nearest spring.

Water was the first thing the new Peasants' Association tackled. A borehole was dug, and a hand pump purchased, but the flow was too slow to supply the village's 3,340 inhabitants without long queues. They decided to tap a hillside spring. Everyone in the village turned out, every day, often sleeping the night at the furthest sections, and by the end of two months they had dug a channel more than 40 kilometres long from the spring to the village square. With the new supply of water they could now make bricks, so they built a whole row of community buildings: offices, court, health post, co-op shop, literacy centre, knitting and sewing room, pottery workshop, kindergarten with two showers, and a ceremonial stand where visiting dignitaries could be welcomed – all painted in horizontal red, yellow and green stripes, Ethiopia's national colours.

But they had taken self-reliance a little too far. They built their water-channel with no outside help or technical guidance, and in 1981 it failed. The hillside soils are deep-cracking clay. Fissures opened in the ground and the water began to drain away before it reached the village. They dug three more shallow wells with hand pumps, but only one produced water. They sank a 130-metre-deep borehole, with a motor pump to lift the water, but delivery was still slow. Every evening in Alesedestu long queues still form.

Since 1983 the village has had three years of unbroken drought. 'Now the community is concentrating on survival rather than on development activities,' explained the energetic Peasants' Association chairman, Mulushewa Gebre Yohannes. After the 1984 season passed with no rain, the villagers set about diverting a river 14 kilometres away onto their land. They built a dam with 30 lorry-loads of stones and earth and dug another channel. The water had to be deducted over a small stream that had carved itself a bed much lower than the level of the fields. They took the tops and bottoms off fifteen oil drums and joined them end to end, propped on poles, to form an aqueduct. In 1985 they were able to irrigate eight hectares, and harvested 24 tonnes of maize, but it was the only harvest they had. In most of Bale the drought broke in 1985, and settlements to the north, south, east and west had good rains – but not Alesedestu. The men ploughed and sowed twice, hoping for rain, but none came. Late in the year a flash flood, carrying a heavy load of silt, broke the diversion dam and scattered the stones downstream.

Chairman Mulushewa tells nostalgically of the days when a hundred ten-tonne trucks used to pull up in the market square to buy up Alesedestu's bumper harvests, and every family had ten or twenty head of cattle. Early in 1986, those days seemed a distant memory. No one was starving. Once a month the Ethiopian Orthodox Church doles out famine relief from two big blue tents next to the kindergarten. Each family gets 46 kilos of wheat – about half what they need. The gap is bridged by wages from labouring on other villages' land, and by selling off tools and livestock. But the pastures, too, are poor, so the cattle are lean and fetch low prices. Only one farmer in ten now has more than ten cattle, many have none at all. The kindergarten no longer has food for the children's midday meal. Classes have been cut to just half the day, and attendance has dropped from 110 children to 46.

I tell this tale because it is important to realize that, in Africa, even the best of projects and the most dynamic organizations may run up against insuperable problems from time to time. That does not mean that they have failed. Alesedestu has been slowed in its onward momentum, but it has not halted. Its determination, if anything, has been steeled and strengthened. The people are still fighting to divert the river, this time by pump and generator, writing to aid agencies in an attempt to raise the $10,000 they need. In 1985 they planted 20,000 tree seedlings. All of them died for lack of water, but they are raising more in a nursery by the river bank, ready for 1986. Even while I was visiting, in January, most of the men were out ploughing on the off-chance of an early rain, though the small rains were not due for another month.

The people of Alesedestu just will not lie down. They have developed an organization, an ability to raise labour in prodigious masses, above all an attitude of mind, that allows them to go on fighting to control their circumstances, despite reverses and temporary defeats. I have little doubt that in the end they will win through.

There are imperfections in the Bale project. The women's income-generating activities have been launched without proper study of their economic and marketing prospects. It seems likely that many of the less economical activities may not survive, as their earnings will not cover purchase of new raw materials or renewal of capital equipment. On the other hand, there is considerable scope for small-scale industrialization if individual villages concentrate on one viable speciality and perfect it, selling their output through the co-op shops.

Another difficulty relates to the limits of UNICEF's brief,

which is to foster the welfare of women and children. That embraces a wide field, from health, education and nutrition to vegetable growing. But it leaves out mainstream agriculture – the basic foundation of health and nutrition, and the ultimate source of all surplus funds with which to pay for other services. Agriculture in Bale has not made the remarkable advances seen in other fields. Until this is remedied, it may prove a constraint.

The third problem relates to Ethiopia's essentially Leninist political structure. The new Ethiopian Workers' Party is seen as the people's vanguard, and the central plan is paramount. But there is a potential conflict between centralized planning and party control, and the ideal of popular participation in deciding what is done at local level. A further danger lies ahead as the Ethiopian Workers' Party expands and consolidates. The peasant association leadership may become politicized and there is a risk that members' wishes, where they contradict government or party policy, may be ignored.

Democracy and self-reliance at local level are not *necessarily* incompatible with the one-party state, but they need a special effort to foster them. If participation and local control are to have any real scope in Africa's one-party states, the central plan must be flexible enough to accommodate them, and the party structure democratic enough to foster them.

THE STRUCTURES OF PARTICIPATION

The Naam movement and the Integrated Basic Services project in Ethiopia have both been highly successful in stimulating a wide range of development activities. These activities, once begun, are self-sustaining. The basic reason is that they are not imposed and managed from outside and above. They are initiated by the people and therefore correspond to their needs and desires. They are founded primarily on self-help, on man-power and resources from within the community. They are structured to be managed by the community. The technologies involved are simplified so that local people, with brief training, can master and maintain them. The target communities have in many cases been transformed into self-developing organisms with impressive capacities, pulling themselves forward across a

very broad front. There were advantageous circumstances in
both cases – in the Yatenga, the existence of a long tradition of
disciplined self-help among the Mossi tribe; in Ethiopia, the
forward momentum created by liberation from serfdom.

But there is hardly any part of Africa where traditions of
mutual help are completely moribund, or where they cannot be
resurrected to form the powerhouse of authentic grassroots
development. What is required for this broad upsurge is a
popular oganization, representing the whole population of each
village, or a large section of it. This body may be what remains
of the traditional community structure. It may be voluntary, or
it may be statutory. What matters is that it should command the
confidence and loyalty of the local population. That means that
its leaders should be elected, not nominated, and the ultimate
decision-making forum should be the assembly of all its members.
Such a body will be able to motivate its members, and to
mobilize them in large numbers for actions that they perceive to
be in their interests.

The organization should possess real powers within its locality.
It should be able to decide on improvements, and to act on its
decisions. For that it must command real resources, whether
mustered from its members, or received from outside. This
process of *empowerment* is the reverse of the removal of local
power by colonialism and the urban-centred modern state. It is
a returning to the base of the rights and command over resources,
a restoration of the dignity and confidence which people need to
feel capable of solving their own problems.

We are not talking here about pure self-help. Self-reliance, in
Africa, must be *assisted*. Most African villages are caught in a
vice between the technology that has stood them in good stead
for so many centuries, and the realities of population and
environment. They need to change or adapt those technologies
as rapidly as possible, and they can only do so with outside
assistance. They also need some basic outside resources: pick-
axes, wheelbarrows, watering cans, cement, hand pumps, or
basic modern medicines. These should, however, be kept to an
unavoidable minimum.

The role of the state and aid donors changes in this new
model. No longer are they the providers of all things related to

development. Instead they become facilitators, providing technical expertise and essential resources to help people to realize their own aspirations, individually and collectively.

Africa's predicament has arisen largely because governments and donors have pursued the Western model of development separately from the rural majority, or sought to impose it on them from above. Once government, donors, and people start to work together in a creative partnership in which both sides teach and learn, Africa will begin to develop in her own authentic way.

PART THREE
The Lessons

When the short road is blocked
you have to take the long road

Mossi proverb

With patience, you can even cook stones

Hausa proverb

17 The secrets of success

Our African journey has been paradoxical. Everywhere we have seen evidence of degradation and desertification, from rainforest clearings invaded by spear grass, through crusted, barren patches where nothing would grow even in years of good rains, to moving mountains of sand amid cropland, and beyond that the majestic but lifeless landscapes of the Sahara, ominous backdrop to Africa's drama.

Yet on the same voyage we have come across spreading signs of hope, areas of advance in a line of general retreat, victories pulled out of defeat.

It is not the purpose of this book to pretend that the battle is well on the way to being won. That is far from the case. By and large the breakthroughs are surrounded and vastly outnumbered by failures. On present trends, disaster will carry the day. But it need not be that way. Our success stories are like seeds. If they are sown widely enough, they can take over the field. In order for them to do that, we have to understand what it is about them that has enabled them to flourish in an environment where most efforts have withered. We have to try to extract the secrets of their success, the factors that they have in common that distinguish them from the failures.

We should first get some common red-herrings out of the way. It is sometimes maintained, for example, that large-scale projects can never succeed in Africa, and that only small-scale efforts should be mounted.

The term 'large scale' confuses a number of distinct characteristics: large in unit size, large in unit cost, and large in spread. Projects involving high costs invested in a small number of huge installations – big dams, factories or irrigation schemes – do have an appalling record. There may be situations where they are unavoidable, but they should only be attempted where the economic returns are overwhelming, and *where there is no low-cost alternative*.

But projects with a wide *reach* are indispensable if any progress is to be made in Africa, and there is no evidence that they are any more liable to failure than geographically restricted efforts. Zimbabwe's programmes for food production and family planning, Kenya's drive for tree-planting and soil conservation, Burkina Faso's 'commando' operations in health, are all large-scale, nationwide, and extremely successful.[1] Of course, when large-scale operations fail, they fail on a large scale, with massive waste of funds, opportunities, and popular confidence. That is an argument for careful planning and preparation of large-scale ventures, with pilot projects to test their feasibility.

A parallel claim is that voluntary projects will succeed where governmental and inter-governmental efforts are likely to fail. The evidence for this is anecdotal. Very few voluntary groups formally evaluate their projects, but *a priori* it does seem likely that their success rate may be higher. Voluntary bodies that are genuine associations of beneficiaries – women's groups, peasant associations, co-operatives – will have a higher success rate, given adequate resources and sound technical guidance, because what they undertake is more likely to correspond to the real needs, abilities and limitations of their members. Foreign charities, too, have certain built-in advantages. As their financial and technical resources are limited, they cannot afford enormous, expensive ventures or complex, imported technologies. They have no choice but to focus on low-cost ventures. Their personnel are low-paid, so more work is achieved for less money. Because of the low pay, only the dedicated apply for jobs, so that staff are often more committed than many an overpaid UN official or career civil servant. Voluntary agencies can more easily circumvent red tape, and are not so rigidly bound by official policy. For the most part they work directly with communities. In many cases a community group will have approached them directly for

---

[1] See index for project references in this chapter.

help, so the project is more likely to correspond to people's felt needs. Charities have no coercive authority, so what they do relies on people's consent and co-operation.

But if Africa is to be saved, efforts on a vast scale are needed, and charities do not have the personnel or the funds required. They have a very important role in pioneering new approaches, and aid agencies should increasingly sub-contract work to them. But only governments and governmental donors have the resources needed for the gigantic task ahead. And these official agents can succeed when they do things right – again as shown by Zimbabwe's food and population programmes, the USAID-funded stove programme in Kenya, or UNICEF's integrated services project in Ethiopia. If they learn the lessons of the African successes, governments and international agencies can succeed too.

Personalities are always important. The quality of leadership, from national down to village level, often makes the difference between success and failure. The people behind the projects and programmes in this book all have an unstinting commitment to what they are doing, a dedication to achieving real benefits for poor people, a resolve to get results. They mean business, and they are not satisfied with the mere appearance of success: they take steps to find out if people are really getting the benefits. Yet they are not aggressive, obsessive types with fixed ideas; on the contrary, they are unusually open-minded, flexible and empirical. Like scientists testing theories, they try things out and watch what happens – if one approach doesn't work, they drop it quickly and try another. Africa is always springing new surprises, as Pliny the Elder remarked. Only those who are ready to abandon all their preconceptions, if need be, have any hope of devising solutions that can work.

Those who succeed in Africa also share a common attitude to Africa's people. They have respect for ordinary African peasants – not only theoretical respect for their human dignity, but practical respect for their views and their wishes, and respect for their accumulated wisdom and traditional practices. It is the converse of the colonial, condescending attitude, still far too prevalent, that the African farmer is incompetent, irresponsible and intent on self-destruction unless suitably instructed in Western ways of doing things.

The projects and programmes we have looked at are pioneering ventures. It is inevitable that the people behind them will be exceptional. But Africa cannot afford to rely on finding large numbers of exceptional characters to solve her problems. Most human beings are fallible and imperfect, and the failings that can be fatal in Africa – stereotyped thinking and jealous insistence on the powers of office – are all too common. Training courses can to some extent instil our project leaders' qualities – openness, flexibility, empiricism and respect for beneficiaries' views and wishes – into ordinary mortals. But these same approaches must also be built into the design of projects and programmes.

PARTICIPATION, EMPIRICISM, FLEXIBILITY

The key is participation. There is not a single one of our projects that does not involve some degree of participation by the beneficiaries. The finding should come as no surprise; almost every major survey of the ingredients of successful projects has pointed to the crucial importance of participation.

Participation should begin from the earliest phases of *project design*. This will ensure that project aims are based as far as possible on people's felt needs, on their priorities, on dealing with what they perceive as important problems. This may not always coincide with what the visiting expert sees as the major problems. For example, villagers may see shortage of poles for building, or wind erosion, or shortage of fodder, as a bigger problem than shortage of fuelwood; tree-planting of various types is called for in these cases, but a project that came in with a purely fuelwood focus would fail.

Participation is also a way of making sure that projects are geared to local circumstances. In the African environment, culture and economy can vary from one village to the next more dramatically than in any other continent. Peasants are *ex officio* the world's greatest experts on their own circumstances. Technologies that seem suitable on the research station have to be tested out under the complex pressures of farmers' lives: their priorities, their daily and seasonal patterns of labour, their command over cash, draught power, manure or other resources, and all the competing demands on their time and inputs. Technologies that look promising at national level may have to be fine-tuned to suit each local level.

Participation in design can take a number of forms, from on-farm research, as with alley cropping, to market research, as in the case of charcoal stoves in Kenya, to village assemblies, as with integrated services in Bale province, Ethiopia. A few of our projects, such as the Majjia valley windbreaks, did not involve participation in design. These succeeded because project managers – whether by accident, by intuition or by an unusual degree of sensitivity and awareness – hit upon approaches that met people's felt needs and served their interests. They succeeded more or less by luck or genius – and that simply is not good enough for general application.

All our projects involve people's participation in the *execution* of projects, in providing land, resources, labour or cash. This kind of involvement serves multiple purposes. It reduces the cost of projects, and therefore allows scarce funds to spread further. It increases people's identification with the project, the feeling that it belongs to them, that they have invested in it, that they have an interest in keeping it going.

Wherever people are paid for their labour contributions, as for cut-off drains in Kenya or soil conservation in Ethiopia, this feeling of identification is weakened. In both these cases payment fostered the attitude that work would only be done if payment was forthcoming, and that maintaining installations built by paid labour was the government's responsibility.

By contrast, when local people were asked to contribute cash to pay village workers' wages, or to buy trees, stoves, wells, pumps and so on, positive attitudes were fostered. When people have to pay for tree seedlings – even a nominal amount – they take a great deal of trouble to make sure that they survive and thrive. They are less likely to abuse equipment they have paid for, and more likely to change other aspects of their behaviour to make sure that they get the full benefit of their investment. For example, people who have invested in a fuel-saving stove often get *higher* fuel savings than in laboratory tests because they take more care in the way they use fuel. When they are paying village workers' salaries, people are more likely to use their services and support their work in other ways.

Payment by villagers allows scarce governmental funds to spread further. In some cases, such as Lutheran World Relief's wells, or the Naam groups' flour mills, payments go into a rolling fund which is used to finance further wells or mills, in theory *ad infinitum*. Payment boosts the dignity of villagers, the feeling of self-reliance. It also serves as a pointer to whether things are suitable or viable in the long term. People will only pay up for things that meet their needs, and things that work. When they stop paying, that is a sure sign that what is being offered should be changed. Free gifts are almost always accepted and then often abused or neglected.

Some reservations must be made. The payment required must always be reasonable and affordable, or it will not be made. Larger payments that promise attractive returns can be covered by loans to be repaid from the proceeds. Payment is always desirable for private goods such as home and farm equipment, tree seedlings and so on, but the case of public goods such as health care, clean water or education is more complex. The old principle 'from each according to his abilities, to each according to his needs' should apply here, but payment by universal contributions is more common. It can also be questioned whether poor farmers should be asked to pay for their own health care, education and so on, as long as government budgets are weighted to providing free health care in clinics and hospitals in cities. Contributions are only equitable where government spending as a whole is equitable.

The third important sphere for participation is in *project management*. The World Bank's Tenth Audit Report found this a crucial factor in long-term project success: 'A major contribution to sustainability came from the development of grassroots organizations, whereby project beneficiaries gradually assumed increasing responsibilities for project activities during implementation, and particularly following completion.'

Self-reliance is often held up as a desirable goal of development: in

Africa it is a pre-condition of success. Reliance on imported techniques and machinery, on foreign finance, on expatriate skills, has led African countries into dependence, and dependence means vulnerability. Projects that involve expensive manufactured equipment and imports or substantial government subsidy will fail at the next foreign exchange or budget crisis.

The same lesson applies at local level: projects that rely heavily on highly skilled manpower from outside the village will start to fail as soon as the expert cannot get through because his jeep is out of action or the road is washed away. *Projects have to be designed so that they are relatively immune to Africa's endemic finance, foreign exchange and skill shortages.* They must make maximum use of easily available local tools and materials. They must rely for the most part on locally raised funds. They must be capable of being managed and maintained to a large extent with local skills – or skills that can be picked up on a brief training course. Their benefits must be so attractive that local people will make spontaneous efforts to keep them going. Projects that meet these conditions will eventually become self-sustaining, capable of continuing and spreading even when government or foreign assistance is interrupted or withdrawn.

As we have seen, successful project leaders have an empirical approach not unlike that of scientists. They are sensitive to, and seek out, feedback on whether what they are doing is effective. If it is not, they adjust it. The market and design research of the improved charcoal stoves in Kenya is a model of this. Other organizations that did not test their stove designs in the furnace of reality failed. Another example is Oxfam's water-conserving stone lines in Burkina Faso: not just the particular technology, but the whole scope of the project was altered when it became obvious that villagers were more interested in food production than in tree-planting.

Projects and programmes should be made flexible, easily changed in the light of information on progress and popular reaction. The gathering of that information should not be left to chance; where possible, basic indicators such as sample crop yields, tree growth rates, infant mortality and so on should be used to assess progress. Village workers or school leavers can be trained to collect these. Even more important, the views of local people on projects and programmes should be gathered regularly so that the beneficiaries become the principal evaluators of project success. If an indicator shows that the expected gains are not being achieved, or if local people feel that something is not going quite right, the project should be altered accordingly. Important services such as agricultural extension should be structured in such a way that they respond to popular demands and pass information up the hierarchy as well as down.

LOW-RISK ENTERPRISE

Successful projects never lose sight of the fundamental facts of life in Africa. African peasants live on, and often below, the threshold of subsistence. Their climate is the most unpredictable of any major region in the world. Their first concern is survival this year, their second, survival next year, and if possible thereafter.

These facts impose certain constraints on development programmes. First, most African farmers have little or no cash or surplus production, and of the little they have, much has to be kept in reserve as insurance against bad rains. Second, they cannot afford to take on board their flimsy life-rafts any additional risk that might sink them. Third, they cannot afford to divert any land or labour from crop production, unless they are virtually guaranteed an additional return that will compensate for any losses. Projects that ignore these constraints are doomed to failure – and what is worse, some of their 'beneficiaries', or at least some of their children, may be doomed as well.

What the African farmer needs, then, is low-cost innovations promising a handsome return for the least possible risk. Most Africans can only afford to gamble on dead certainties. *None of our successful projects involves any increase in exposure to climatic risk; indeed, many of them reduce risks and protect farmers and their crops and animals against the vagaries of the climate.* Soil-conservation work, for example, increases the amount of water that filters into the soil, and therefore makes the most even of poor rains. Windbreaks reduce loss of moisture through evaporation, and reduce the burying of seedlings by sand. Fodder trees reach subsoil waters and produce some food for livestock even in drought years.

Most of our successful projects are very low-cost or no-cost in terms of cash up front demanded from the beneficiary. Zimbabwe's package of fertilizers and improved seeds is relatively expensive – but the goods are provided on credit, to be repaid out of the handsome harvest surplus. No surplus, no repayment.

Labour is often in short supply in Africa, especially around planting and harvesting time when a day's delay can sometimes mean a significant loss of yield. Therefore *our successful projects are also low-cost or nil-cost in terms of labour demand at peak times.* By and large they use labour in the dry season, when it will not interfere with food production. Where they do involve labour in the growing period, as with agroforestry, they reward that labour with a worthwhile additional return in fruit, fodder, or fuel.

Another category of costs is often overlooked: the benefits forgone. Land may have to be taken out of food production to make room for terraces, windbreaks or tree plantations. Grazing may have to be limited to give trees a chance to get going. These costs are just as critical as cash costs to the farmer who depends on every last kilogram of grain, meat or milk. *The most successful projects and technologies minimize the costs in forgone benefits, and more than compensate through raised crop yields or other products.* Oxfam's lines of stones in Burkina Faso, for example, take up no more than 5 per cent of the land in exchange for a 50 per cent increase in crop yields. Alley cropping takes perhaps 5 per cent of the land, but raises crop yields by 30–50 per cent and provides fodder and wood in addition. The greater the forgone benefits, and the lower the compensating returns, the greater is the risk that the farmer will not adopt the required change, or will quickly drop it. Stone terracing is a case in point.

All our projects represent very attractive investments for peasant beneficiaries. *The pay-off period for productive investments is almost always a single year, sometimes substantially less.* Kenya's improved charcoal jikos pay for themselves in less than three months; the diesel pumps for small-scale irrigation in Kano within one season. The labour spent on building the improved clay stove in Burkina Faso is saved in two or three weeks. Trees take a couple of years to pay off in humid areas, longer in semi-arid areas, but provided the cost in forgone benefits is low, and farmers know that they will reap the eventual products, the longer pay-off period is tolerated because people can see the potential wood or timber production build up year by year.

In World Bank projects, returns above 10 per cent are considered acceptable. Few of our projects have calculated the *rate of return to participants – but in most cases these would be of the order of 50–100 per cent a year, sometimes much higher*. Returns of this sort are what is needed to attract subsistence farmers exposed to severe climatic risks. Once their confidence is gained and their incomes increased, they may adopt innovations with a lower return, but if they are involved even once in a venture that costs them their surplus, or a child, they may never trust a government employee again.

For non-productive investment like health, similar considerations apply. Child-spacing, prolonged breast-feeding and oral rehydration therapy are all virtually cost-free to users, yet promise reductions in infant and child mortality of the order of 40 per cent. Prolonged breast-feeding has the added advantage that it is the lowest-cost, most agreeable and straightforward form of contraception in existence.

Low unit cost to governments is another important factor in success. Low-cost ventures are much more likely to achieve a satisfactory return, because even quite modest improvements in output or welfare will justify a small investment. Low initial costs mean that the same investment can be spread much further. Except in a handful of countries with secure, diversified exports or long-lived oil reserves, *projects should be designed with low recurrent costs*. This makes it far more likely that projects and programmes will be sustained through the repeated budget crises that are likely to remain a fact of life in most of Africa for the foreseeable future. There are many ways of keeping costs low. They include maximizing popular participation in carrying out and sustaining projects, and the use of barefoot village workers of all kinds, from peasant conservation cadres to family planning distributors, health agents and nursery teachers. *It is important to keep the use of imported inputs such as machinery or chemicals as low as possible.* Imports become vulnerable just as soon as the foreign aid backing the project dries up. Very few of our successful projects are dependent on imports. Those that are, even when the imports are food aid, as with Ethiopia's conservation programme, are to that extent vulnerable.

SPREADING THE MESSAGE

The technologies launched by some of our projects were so outstandingly successful that they began to spread by themselves: the improved cassava varieties of the International Institute for Tropical Agriculture in Ibadan; the use of phosphate fertilizers in Niger; high-yielding cowpea seeds in Kano State, Nigeria, to mention but a few. They needed no advertisement other than their own performance, and no persuasion for peasants other than the evidence of their own eyes.

But most innovations – including these three – spread faster if they are actively pushed on a nationwide basis. In many countries an informal network for spreading innovations already exists: the market. Even in the poorest countries, periodic rural markets reach almost all peasants at certain times in the year – especially after the harvest, or in the 'hungry gap' when cereals are bought in. *In most countries, the market is the best channel for disseminating movable items for the individual household, from metal stoves, maize shellers and hand grinders, to improved tools for the farm.*

In social marketing, government or aid agencies act as the venture capitalists in launching new socially useful products, financing research, training and advertising. Kenya's charcoal stoves programme was one of the first projects to recognize and use the full potential of the market. It succeeded because it accepted all the disciplines of the market, particularly the need to offer a product that was acceptable to consumers and gave real benefits for an affordable price. Thorough consumer research, and use of artisans and dealers involved in the less efficient older technology, were essential ingredients.

Technologies that are not movable, or that involve ways of doing things rather than hardware, demand different channels.

In health, education and agriculture, nationwide networks of grassroots workers reaching into every district are essential. Zimbabwe's drive for increased food production and Kenya's soil conservation programme are both mounted by nationwide extension services.

In both cases *the productivity of extension workers is increased wherever they work with groups, rather than individual 'model' farmers.* Community groups were active in several other successful programmes, including the spread of stone lines among the Naam groups of Burkina Faso, mobilization for vaccination by Burkina's Committees for the Defence of the Revolution, or broad-based development initiatives among the peasant associations in Ethiopia's Bale province. *Groups speed up the process of innovation* (provided it fits our other criteria) *by social influence and emulation. They spread the net of personal contact far wider.* They act like adult education classes for mutual instruction. They can also serve as an effective source of feedback so that techniques can be gradually improved and adapted to local circumstances.

Training is central to rapid dissemination. Shortage of qualified manpower has always been considered a block to development in Africa. Using conventional methods, training takes a long time and often barely keeps pace with wastage to the private sector, or promotions. *Yet national networks of thousands of village-level workers can be put in place within a year.* Burkina trained 7,000 village health agents and 7,000 traditional midwives within six months of taking the decision. The secret was the pyramidal training system, in which, for example, the skills of five national-level experts can train fifty to a hundred regional trainers, each of whom will train twenty further trainers, or grassroots workers. *The content of courses must be simplified to the basic essentials needed to have some valuable impact, and in most cases can be done within two months – with refresher courses in subsequent years.* In most cases – as in Burkina itself – the villages that will benefit may be willing to pay the travel and subsistence costs of the village worker. *Using the pyramid approach, one expert's most relevant knowledge can be multiplied twenty-fold in two months, four hundred-fold in four months, and eight thousand-fold in six months.*

The messages carried by these purpose-built systems must be backed up by reinforcing messages carried on every available channel of information. Like agroforestry in Kenya, they can be made part of the curriculum in schools, and used as practice reading material in literacy classes. Like oral rehydration therapy in Gambia or vaccination in Burkina Faso, they can be broadcast on TV and radio, through rural broadsheets and travelling popular theatre groups.

The techniques and technologies have to be designed with ease of dissemination in mind. Oxfam's stone lines and the water level for laying them out were refined until they were so simple people could master them in a couple of days. The basic principles of alley cropping are so straightforward they can be put across in six simple drawings with one-sentence captions. The design of the improved jiko in Kenya took into account the ease with which it could be mastered by market metalsmiths. The home-made formula for oral rehydration salts – a pinch of salt, a cupped palmful of sugar in a glass of boiled water – can be picked up in ten minutes. Techniques that are partly based on familiar approaches will be more quickly understood and adopted: for example, many parts of Africa have some existing form of agroforestry or some traditional method of soil conser-vation. Conservation will make faster headway if it builds on these existing systems and develops them, rather than attempting to introduce entirely new and unfamiliar techniques. For the same reason family planning can make much better progress with the idea of child-spacing, which formalizes a common African practice, than with the novel Western idea of limiting family size.

That said, African peasants will quickly adopt completely unfamiliar technologies where these are affordable and dramati-cally beneficial and do not involve radical changes in other important aspects of life. One need only think of cassava or eucalyptus, bicycles or antibiotics. It takes a year or two longer – they will usually want to see it working in a neighbour's field or home, or in a small corner of their own, before they adopt it whole-heartedly. African peasants are sceptical about innovation – it is a wise part of their strategy to avoid risks. But they are extremely fast adapters once the evidence of their own eyes has convinced them.

THE POLITICAL CONTEXT

The political context can interfere with the best-designed ventures. Real participation may be blocked by governments opposed to any expression of popular autonomy – or by bureaucracies and ruling parties with vested interests in maintaining the power of officials and technocrats. The economics of development may be distorted by pricing policies that favour cities over rural areas. Even low-cost, self-help projects can be crippled when the modest resources or technical assistance they need are lacking.

We have seen successful projects under the whole spectrum of political alignments and systems, in multi-party states and single-party states, in representative democracies and military regimes, in capitalist and in communist countries. Success is inherently possible under any of these systems; except in the case of the very few totally corrupt and brutal regimes, and the racist apartheid regime of South Africa and Namibia, the overthrow of any particular political system in favour of any other is not an indispensable pre-condition of progress.

What *is* a pre-condition, however, is genuine and effective government commitment to improving the welfare of the rural majority, and willingness to create the conditions for community and individual initiative. A government that insists that it alone knows what is best for each community, and that its officials alone can deliver services or bring about change, is certain to fail in almost everything it does. By contrast, when a government gives whole-hearted support and technical guidance to well-aimed programmes with strong local participation, almost anything is possible. The backing must be coherent and continuous. Verbal commitment is important – but clearly not enough. Kenya's President Moi, for example, frequently speaks out on the need for reduced birth rates, yet Kenya has not mounted a well-funded, well-organized family planning programme. Actions speak louder than words. The touchstone is commitment of significant resources, in funds and manpower. The commitment must be consistent: all government departments must be pulling in the same direction, in a co-ordinated way. For example, it is no use conservation workers urging farmers to intercrop or to plant trees, if agricultural extension workers are advising monocropping and the removal of trees to allow a straight run for ploughing. Co-ordinating mechanisms have to be set up to ensure consistency – like Kenya's Presidential Commission on Soil and Water Conservation and Afforestation.

Perhaps the most important thing any government can do is to create the conditions in which people can develop themselves. Price policy is central in the areas of agriculture and environment. As the experience of Zimbabwe shows, attractive prices are essential if farmers are to have any incentive to increase their food production. Rising milk prices stimulated the spread of dairy cattle among Kenya's smallholders. Rising wood prices are stimulating the nationwide wave of tree-planting in Kenya. *Without adequate prices, no other action to increase food output or tree-planting will succeed*. With adequate prices, farmers will develop many of their own ways of intensifying production, and the impact of government actions will be multiplied. Anything that stands in the way of farm prices rising to the level necessary to meet demand – from regulated prices, or government marketing monopolies with low purchasing prices, to cheap imports – will stand in the way of national self-sufficiency in food or in fuel.

Where conservation or production demand longer-term investments in the land, such as tree-planting or terracing, farmers must have the power to carry out works, and the certainty that they, or their children, will enjoy the benefits. One reason that soil conservation and reforestation have been so successful in Kenya is because most farms are privately owned, and owner-occupied. Farmers can do what they like on their own land, and they know that they and their children will enjoy the full benefits of any improvements they make. Any form of land tenure that reduces the farmers' freedom or incentive to carry out long-term investments works against conservation. This includes communal ownership with shifting cultivation, landlordism, and state ownership. Even within these systems, farmers can be given lifetime security of tenure of their farms, rights to plant trees, rights to pass on the property to their children.

However, the length of security needed depends to some extent on the speed with which an investment pays back. For terracing on steep land to be worthwhile, at least fifteen years' security would be needed; for stone lines on almost flat land, five years would be enough; for tree-planting, the length of time needed for the tree to reach harvestable size – three years or less in humid areas, up to ten in semi-arid. In general, *the faster the pay-back period for an innovation, the less of an obstacle does lack of secure tenure pose.*

Donors are always complaining of lack of government commitment to projects, but donor commitment is just as important. Donors, too, must be consistent and continuous in their support. Most projects run for a maximum of five years, the usual horizon of most elected Western governments, but entirely inadequate for building nationwide programmes to halt deforestation and land degradation, to improve health or family planning services. As we saw in Chapter 3, all too many aid projects do their best to include a hefty chunk of imports, and donors are reluctant to shoulder local and recurrent costs. As a result, projects are made too dependent on imports, and assume continued budget support from African governments. As soon as the aid funds are withdrawn, the project is wide open to the budget and balance of payments crises that almost always materialize in Africa sooner or later. As we have seen, donors are often fickle and primadonna-like. Aid fashions fluctuate, and donors push their own separate line, often in competition with each other.

All these practices will have to change. Our two most successful soil conservation ventures – by the Swedish International Development Association in Kenya and the World Food Programme in Ethiopia – are models of responsible donor behaviour. *They have been running for a decade and are still going strong.* In both cases *imports of hard-to-repair machinery are limited only to transport vehicles. Local and recurrent costs are paid.* The governments can plan for the long haul, assured of consistent support. *In each case a single donor dominates the field concerned, and has single-mindedly pursued the same objective without wavering.* Neither project has created favoured regional enclaves with their own separate resources and staff; *both have worked on a national scale, through national government departments, aiming to strengthen national capacities.* Donors who genuinely wish to help Africa should follow suit. Conversely, African governments press for aid under these conditions. There is a case for every African country to create an inter-ministerial commission for agriculture and conservation, to co-ordinate aid as well as internal policy. Multiple donor styles or approaches should be unified so that a single approach dominates in each sector.

Any government or donor that followed these guidelines could hope to meet with a similar measure of success as that achieved by the projects described in this book. To those whose goal remains to modernize and industrialize their nations as speedily as possible, the approach may seem basic and unglamorous. What we are dealing with here are foundations and first steps. Only if the first steps are sound, and well-directed, is there any hope of further progress. Only if the foundations are secure can anything be built on them that has a hope of lasting.

THE KEYS TO SUCCESS – A CHECKLIST OF BEST PRACTICE

Costs to beneficiaries
- Low or nil cash cost (alternative: credit programme)
- Low or nil forgone benefits – especially little or no loss of food production
- Low labour input at crucial crop periods
- No increase in exposure to climate risks

Benefits
- Financial rate of return 50–100 per cent or more
- Pay-back period one year or less
- Reduced exposure to climate risk

Easy Local Maintenance
- Low import content
- High content of locally available materials
- Low skill requirements of maintenance

Easy to Disseminate
- Based partly on familiar principles
- Teachable in 1–2 day courses

Mass Dissemination Approach
- Effective nationwide network with mass contacts:
  - Market
  - Extension system
  - Mass organizations
- Pyramid training
- Strong emphasis on education and awareness:
  - Use of all communication channels, schools, media, bureaucracy

Strong Local Participation
- In design via:
  - on-farm research
  - market research
  - social surveys
  - popular assembly

- In execution via:
  - contributions of land, labour, materials
  - some payment – nominal, or on credit
- In management and evaluation:
  - aim: self-sustaining
    self-financing ventures

Learning Process Approach
- Regular feedback on performance via:
  - basic indicators
  - contacts with beneficiaries
- Flexibility:
  - open to continual adjustment in view of performance, obstacles, opportunities
- Pilot projects to test and perfect untried approaches

Political Backing
- Verbal and symbolic
- Resources and manpower
- Policy environment:
  - favourable food and wood prices
  - security of tenure of land, trees, pastures, water points
- Consistency and coherence of all government acts

Donor Backing
- Ten-year horizon
- Payment of local and recurrent costs
- Consistency of goals
- Co-ordination with other donors:
  - Unification of approaches within each sector
- Commitment to strengthening national capacities
- Sub-contracting to voluntary organizations

18 The prospects for a quantum leap

Forecasts are always dangerous. Human societies are too complex for the future to be predicted with any degree of confidence: the element of free will confounds all attempts at clairvoyance. People are not automata determined by their circumstances: they can at any time change course and alter those circumstances. From victims of their environment, they can become masters. That is true of individuals – and it is true of whole societies.

It is safer to talk of scenarios. For simplicity's sake, in this final chapter, I shall look at just three alternatives that could lie ahead for Africa. The first is the one we hear most about: the future as the linear continuation of the past, the logical outcome of present trends. That way lies probable catastrophe.

In the second scenario, certain self-correcting mechanisms come gradually into play. They may stave off total collapse, but do not avoid stagnation and continued poverty.

In the last scenario, a swift rupture is made with past trends through a shift into a new set of relations between people and a fragile environment which can form a stable basis for progress. That way lies hope.

THE DOOMSDAY SCENARIO

The most obvious scenario is that the trends of the past decade and a half will persist. Population is projected to grow at around 3 per cent a year until the end of the century. The population of Africa as a whole is projected to grow from 553 million in 1985 to 887 million in the year 2000 and a staggering 1,643 million twenty-five years later. It may eventually climb to an almost inconceivable 2,500 million – four and a half times the present level – before flattening out. The problems of rapid population growth already seem insufferable for Africa – but the worst is still to come. Between 1975 and 1985 an extra 14 million Africans were born each year. At the peak growth of absolute numbers, in the years 2010–2015, the annual additions will be no less than 32 million a year – more than double the present increments. This is equivalent to adding an extra Nigeria – the largest country in Africa – every three years.

Food production per person will continue to decline at around 1 per cent a year. The Food and Agriculture Organization has projected the outcome of these trends in its study *Agriculture: Toward 2000*. In 1975–9, Africa – once self-sufficient in food – produced only 83 per cent of her cereal requirements, and imported 8 million tonnes to fill the gap. By the end of the century, net cereal imports will have risen sixfold, to 49 million tonnes, and Africa will be growing only 56 per cent of the cereals she needs. Many countries, still crippled with balance of payments problems, poor terms of trade and debt, will be unable to import what they need, and many poor Africans will be unable to afford to buy what they need even if their country as a whole has enough. As a result, the numbers of severely malnourished people will rise by more than three quarters, from 72 million in 1974–6 to 127 million in the year 2000.

The food crisis will be paralleled by an equally acute shortage of fuelwood. The numbers of people short of fuelwood will rise from 180 million to 290 million. In 1980, in sub-Saharan Africa, there was a rough balance of fuelwood supply and demand (though of course, many local situations were critical). By 2000 A.D. there will be a shortfall of 220 million cubic metres.

After the year 2000, the potential conflict between population and the resource base will grow more and more acute almost everywhere, except in the humid zones of central and central-southern Africa. By the year 2010, eight countries will be using 90–100 per cent of their cultivable land: Niger, Togo, Somalia, Rwanda, Burundi, Kenya, Botswana and Lesotho. Another five – Nigeria, Sierra Leone, Gambia, Mauritania and Malawi – will be farming more than three quarters of their cultivable land.

FAO's study of the carrying capacity of land in developing

countries (p. 243) compared Africa's projected future population
with the food production potential of her lands. The number of
countries that would be unable to feed themselves from home
production using the present low level of inputs would rise from
22 out of 49 in 1975 to 32 by the end of the century and 35 by
2025 A.D. Indeed, even as early as 2000 A.D., 16 countries would
be critical even if they used intermediate inputs. They include the
five North African countries, plus Mauritania, Niger, Somalia,
Rwanda, Burundi, Kenya, Lesotho, Namibia and the Indian
Ocean islands Mauritius, Reunion and the Comoros. Nigeria
and Ethiopia would be close to being critical. As we saw (p. 245),
these results are based on the use of every scrap of suitable and
marginal land to grow nothing but food crops. If we consider
only the production from land that is likely to be actually
cultivated, and deduct one third for non-food crops and unequal
food distribution, the results are more alarming. Then, by 2000
A.D., Africa would be able to feed only 55 per cent of its
population with low inputs. By 2025, it could feed only 40 per
cent.

Most countries of the Sahel and mountainous East Africa will
face severe problems. Ethiopia's 1983 population of 36 million
will more than treble to 112 million in 2025 – 44 million more
than it could feed with intermediate inputs. Nigeria's population
of 2025 is projected to reach 338 million, 123 million in excess of
its carrying capacity with intermediate inputs of 215 million.
Even with high inputs, Kenya's lands could support only 51
million people – a total that will be passed by 2010. By 2025
there may be 83 million Kenyans, with as many as 111 million
before the population reaches its plateau.

Central Africa will face no land shortage even if it is still using
low inputs. With intermediate inputs, Zaire alone could feed
1,280 million people – 95 per cent of the projected total for the
whole of sub-Saharan Africa in the year 2025. But this enormous
surplus capacity is based on clearing most of the rainforest for
agriculture. Even if this could or should be done, it would not
solve the food problems of the Sahel or East Africa any more
than North America's present surpluses can. The only way this
huge potential could alleviate population pressure elsewhere is if

there were massive migration into Central Africa from the surrounding areas – a solution which is fraught with political and environmental problems.

The facile response to these prospects is to point out that many *developed* countries are not self-sufficient in food or energy: they pay for their imports with exports of manufactures or services. Could not Africa's food deficit countries do the same? The problem here is that the prospects of industrialization in Africa are dimmer than in any other region. Apart from oil-rich countries and a few city-states deluged with foreign investment, no country has industrialized without a reasonably healthy agricultural base. Africa seems to be caught in a trap: industrialization could solve her food problems, but with agriculture stagnant, the chances of industrialization are slim.

These grim prospects are all based on the assumption that past trends continue. They are entirely realistic, and they point towards disaster: Africa as the world's nightmare, a continent of recurrent drought and famine and bloody warfare, perpetually dependent on aid and food hand-outs, with spreading deserts and shrinking forests. It could all too easily happen.

SELF-CORRECTING MECHANISMS

But catastrophes are unpredictable things. There are sudden discontinuities when things shift on to different planes, which can be lower – or higher.

There are certain pointers which suggest that past trends may not, in practice, run their full course, ending in continent-wide collapse and devastation. It seems more likely that self-correcting mechanisms will come into play.

The relationship between people and nature in most of Africa has been extractive: people took crops, and wood, and put nothing back, but gave nature time and space to restore herself. That relationship depended on an abundance of land and forests. But as population density grew, that abundance could no longer be taken for granted. Nature is no longer given time to restore herself. Her capital of resources is being depleted, and the whole system grinds gradually downhill.

Everywhere else, where population densities passed a certain point, the relationship between people and nature changed. Farmers are forced to shift gradually from extensive methods – moving on to new areas when old ones are exhausted – to intensive ones, working harder to maintain the fertility of existing areas. This shift has been under way for a long time in the most densely populated parts of Africa, and as other areas fill up, it must happen there too.

The combination of increased population density and loss of land to degradation means that there will be less and less land per person. It will become economically more attractive to invest more labour in each piece of land, to conserve existing farmland, and even to rehabilitate lost and degraded land. The enthusiasm with which Nigerian farmers have taken up valley-bottom farming, or the Mossi of Burkina Faso the water-conserving stone lines, are signs that this *is* happening.

As densities increase, fallow areas decline and the area of open grazing in settled areas decreases. More and more livestock owners will be forced to keep their animals at home and stall-feed them, so the pressure on vegetation will be lessened.

The growing shortage of fuelwood will first create a market in wood, and then force prices up, to the point where growing trees for sale will become a profitable venture. At this point the supply of wood will begin to increase again. This is clearly happening in Niger's Majjia valley, throughout Kenya and elsewhere.

As landholdings are divided and sub-divided at each new generation, they dwindle in size, so farmers do not need as many children to help them out; at the same time they simply cannot feed as many children as their parents could from holdings twice as big. It is the availability of extra land that makes Africans so prolific: when the frontier closes up, the birth rate too may well fall. There are signs that this is beginning to happen in Zimbabwe, though the high birth rates of overcrowded Kenya and Rwanda suggest that holdings may have to shrink to less than two acres before any effects become apparent.

These processes define a probable lower limit to the cycles of decline that are under way in Africa. They may come into operation to prevent the worst-case scenarios from happening, barring major climatic disasters. On the other hand we cannot draw much comfort from this. The end result is peasants working much harder than they do now, to wrest higher yields from the same area of land, but probably with steadily declining incomes per head and still no agricultural surplus to fuel an industrial revolution.

Gradual intensification may well end up by changing the agricultural system, from the extensive, extractive form of shifting cultivation, to intensive settled farming. But it will do so in the African environment only after a great deal of damage has been done. Topsoil will have been irrevocably lost, lowering yield potentials, and much land that could have grown crops will have degraded to waste or pasture.

THE CONDITIONS FOR RADICAL CHANGE

There is an alternative. It requires a shift to soil-conserving, intensified agriculture *before* the mechanism of land shortage brings it about; a shift while individual landholdings are still large enough to be able to produce a surplus. A parallel change is needed in the context that determines the number of children people want: a move to universal education, literacy and health care, focusing especially on mothers and children.

This is precisely the right moment at which any African government with the will could begin such a shift. The technologies with which to do it already exist, or will soon do so. Low-cost techniques of soil and water conservation have been developed which can not only sustain but increase food production. Improved varieties of maize, cassava and cowpea are spreading. Promising lines of sorghum have been identified and have already been released in some countries, though a breakthrough in millet is further away. Africa could embark on its own Green Revolution within the next decade, provided governments are committed to spreading the new technologies. In the field of human resources, a number of high-impact, low-cost interventions have been perfected. The radical simplification of the message, plus pyramidal training schemes – trainers training trainers of trainers (p. 310) – can accelerate the pace at which any technology can spread, once its effectiveness and economic attractiveness have been proven.

Two severe droughts within a decade have created the psychological conditions for radical change. These droughts have had a profound effect on many of the peasants I spoke to. They are much more aware of the impact their own practices may have had in degrading the environment. Many of them are surrounded by the visible evidence of spreading and deepening decay and damage, and the effects are incised into their daily lives through lowered food production and fuelwood availability. Repeated drought and degradation have been brutal teachers but the lessons *are* being learned, by farmers and governments alike. In many areas there has been a shift from a passive to an active attitude to the environment. The natural regeneration on which Africans relied in the past can no longer be taken for granted. Having realized that people can severely damage their environment, they are also realizing that people can act to protect their environment.

Africa's financial crisis, paradoxically, provides the third window of opportunity. High debt repayments, and the adjustment programmes that go with them, mean that African governments will find themselves, over the next five years, with even more constrained budgets and more limited foreign exchange than in the past. If they persist with the old style of development project, involving high import content and high government spending, they will be forced to reduce the number of projects drastically.

But they can seize the moment to change the style; to shift the emphasis towards low-cost projects, with low-import content, that replace much of the central spending with local self-reliance. As we saw in the last chapter, projects of this kind are notably more successful than the old style. This change of gear will allow limited government spending and foreign exchange to spread much further, and to be used far more efficiently. The same amount of investment will produce a far better return.

SHIFTING THE BIAS

The required strategy is one of *going for the quickest gains first: focusing on those actions that produce the biggest impact for the least cost.*

The very low starting point of smallholder agriculture makes it perhaps the most promising area for action. Because yields are so abysmal, and inputs so minimal, gains from modest investments can be very dramatic, with yield increases of 20 per cent upwards, and financial returns of 50 per cent or more.

As we have seen, until recently West African governments neglected small farmers in favour of the urban sector. The problem was seen as a zero-sum game: if cities and industry were to develop, the resources had to be taken from the rural and farming sector, by way of taxes and low procurement prices for food and cash crops. The strategy produced nothing but low output, dependence, mass poverty and tiny, stagnant home markets for industry. Everyone lost out: it became a minus-sum game.

In a continent where seven out of ten people are farmers and their families, healthy growth is possible only where agriculture prospers. If agriculture is fostered rather than exploited, the situation can shift into a plus-sum game – one in which everyone gains. With higher producer prices, farmers have an incentive to grow more. They can earn enough to buy manufactures, and the market for local industry expands. Industry develops in closer harmony with agriculture, servicing farmers' needs for tools and basic household items, and processing agricultural produce. Agriculture can become the shunter getting the whole economy on the move.

Small farmers must be the focus of all efforts. They constitute the overwhelming majority of farmers in all African countries, and because land is still fairly equally distributed, they account for most of the farmed area. They also make up the majority population, and the majority of the absolutely poor and malnourished. *Hence, in Africa, helping the small farmer achieves, simultaneously, four of the most important goals of development. It is the best way of boosting national self-sufficiency in food, the only way of improving the incomes of the majority and of reducing mass poverty and malnutrition, and the only way of conserving the environment of which the African peasant is, for better or for worse, the effective custodian.*

Prices are the key. Where governments control the marketing of food crops, the prices paid to farmers must be increased at least to the level of world market prices, or higher, as long as this does not produce an unsaleable surplus. To avoid overburdening tight budgets, the indiscriminate subsidies that keep food prices cheap for all city dwellers should be gradually phased out except for low-income-group foods like cassava, millet or sorghum. Where government agencies have a monopoly in buying cash crops, they should pass on a far higher proportion of their receipts to the farmer, instead of siphoning off a high percentage as disguised taxation. Ideally these marketing monopolies should be abolished to allow prices to find their own level and to allow farmers to seek the most attractive outlets for their produce. Access roads will be needed to get produce to market, and inputs to farmers. These can be built and maintained by low-cost, labour-intensive methods, creating dry-season jobs close to home.

Exchange rate policies must be altered, too. Overvalued currencies must be devalued significantly, or best of all allowed to find their own level. This will help to boost home agricultural production by making food imports dearer, and increasing the value, in local currencies, of cash crop exports. It will allow governments to abolish the whole paraphernalia of currency controls which take up so much scarce manpower and offer such wide scope for corruption. Higher import prices would keep the level of imports down, so import controls could be loosened or scrapped altogether.

These policies can produce dramatic and rapid results. We have already seen the impact in Zimbabwe (p. 87). A number of other governments, including Mauritania, Mali, Madagascar, Zaire, Guinea and Cameroon, have followed similar policies in the last few years. Zambia devalued in 1984 and increased the official price of maize by 35 per cent: in 1984–5 maize production rose by more than 20 per cent over 1983, and the amount of maize marketed rose by an estimated 55 per cent. For the first time since 1976, Zambia came close to meeting her home requirements. Ghana made a hefty devaluation in 1984–5, and raised the maize price by 300 per cent: maize production tripled in 1984, and was two thirds up on pre-drought levels.

The priority in agriculture should be to focus on food rather than cash crops. This will reduce dependence on food imports, and increase food security by making food available where it is needed. It will also improve the distribution of food, helping to solve the nutrition problems of small farmers. Since men control the proceeds of cash crop sales, and do not always use them wisely, focusing on increased food production will improve the welfare of women and children.

On the other hand, it is important to keep a balanced view on the cash crop issue. It is perfectly possible for farmers to increase their output of food and of cash crops at the same time. The three African countries with the best record in *sustainable* increases in food production – Swaziland, Malawi and Rwanda – did precisely that. The crops to focus on are those with good price prospects, especially those that are not competing with highly subsidized Western production (such as tea, coffee, cocoa) or that may be sold out of season in Europe (like market garden produce). A number of cash crops are ecologically useful, helping to conserve soil and increase production. These include legumes like groundnuts or soybeans, grown in rotation or intercropped with cereals. Tree and shrub crops like coffee, cocoa, bananas, oil palm or coconuts can be used in multi-storey agroforestry, with food crops on the ground layers. A close carpet of tea bushes provides excellent ground cover against erosive rainfall.

CONSERVATION FOR INCREASED OUTPUT

Increasing prices paid to farmers will almost always raise output. But there is no guarantee that the rise will be sustainable. Farmers may simply expand the area they cultivate and invest more labour, without increasing their use of fertilizers or improved seeds, or their soil conservation efforts. If this happens, increased production may mean even shorter fallow periods and a speeding up of deforestation and erosion.

Because of Africa's extremely sensitive environment, and the massive increase in population it will have to support, it is crucial that increases in food production should be *sustainable*: they should not be achieved at the cost of undermining the basis of future food production. Conservation is an indispensable condition of continued production.

But the economics of conservation are tricky. When development agencies like the World Bank are calculating economic returns on projects, conservation normally rates low. Its benefits are mainly expected in the long term, after the project is over. Such future values are normally included in calculations in terms of their present value: that is, the cash amount that would build up to the future value, over the period involved, at the assumed rate of interest (the discount rate). This procedure means that the total destruction of the soil base over a forty-year period would count for very little against a substantial five-year increase in output. *There is no discount rate low enough to give long-term conservation its crucial importance.* Conservation needs to be built in, instead, as an absolute pre-condition of all projects.

But it is not only development agencies that find conservation economics tricky. African peasants do too. Over most of the continent they do not enjoy security of tenure: they cannot be sure that they, or their children, will still be allotted the same area in twenty years' time, to enjoy the benefits of their conservation efforts. This insecurity applies in areas with communal landholding, with state-owned land or insecure tenanted land. It applies to most rangelands, where government-provided wells mean that pastoralists can no longer control access. Insecurity of tenure of trees and forests is also widespread, and discourages planting and sensible management.

African farmers and herders must be granted security of tenure and rights to control land, trees, water points and rangeland, before they will begin to think in terms of long-term conservation. This security will often take the form – as in Kenya – of full private ownership with all normal rights of inheritance, renting and sale. But lifetime security – and the right to pass a holding on to children – can be granted even under systems of state and communal landownership. Women must be given equal rights in any reforms. Tree tenure must be changed to give individuals and communities clear and recognized rights to plant and harvest trees and to use and manage natural forest. In rangelands, specified clans or pastoral associations can be granted rights to control access to certain areas.

Farmers who do not enjoy security of tenure have a planning horizon of only five to ten years. Where drought risk is high, the horizon may extend only one or two years ahead. Under such conditions, the conventional approach to conservation will not work: appeals to preserve the land for the future will fall on deaf ears. In these circumstances, conservation can only be successfully spread if it serves other important goals, like making firewood available closer to home, or providing more fodder for livestock. Fortunately, as we have seen, most conservation measures conserve plant nutrients and water as well as soil.

Conservation can and must be promoted as one of the most cost-effective ways of increasing overall production, and of reducing vulnerability to drought and dry spells. In areas that are susceptible to drought, or to dry spells in the growing period, conservation methods should be promoted that reduce the vulnerability to drought – methods that store water, such as small dams; methods that increase the amount of water that filters into the soil, like water-harvesting and water-spreading; methods that increase the water-holding capacity of the soil, like the use of crop residues, composts and manures.

Because of Africa's widespread labour shortage, methods that minimize the use of labour – or that provide high returns for additional labour – stand the best chance. Managing natural forest involves less labour than planting a village woodlot. Keeping livestock off a hillside to allow trees to seed themselves and grow is easier than replanting the hillside with seedlings. Infiltration zones, grassed strips or contour hedges are less laborious – and more productive – than terraces.

A new style of thinking is needed. Production and conservation should no longer be separated. *Specialists in crops, livestock and forestry production should 'think conservation' and look for ways of increasing production that also improve conservation. At the same time conservation experts should 'think production', and concentrate on methods that improve output of food, fodder or fuel and conserve water for the least cost in cash, land or labour.* Bunds and terraces should never be seen simply as conservation structures, but as areas where useful plants can be grown.

New ways of thinking demand new forms of organization. African peasants operate complex systems which nearly always involve crops, livestock and trees together. Government services must be at least as sophisticated as the clients they serve. Departments of crop production, livestock, forestry and conservation should all be integrated in super-ministries. Obviously there will still be specialists in each discipline, but they will be trained or re-trained to be aware of their close interdependence.

In national and international research, farming systems departments will serve a co-ordinating role, making sure that specialists' work is appropriate to the real circumstances in which it will be applied. Technologies and packages will be pre-tested in collaboration with local farmers in a wide range of areas, perfected and simplified so that they can be taught quickly.

A national network of extension workers reaching into every village will help to disseminate the new approaches. These can be set up very rapidly using the pyramidal training approach (p. 310). At village level there will be a single multi-purpose agricultural agent, giving advice on all subjects, paying attention to trees, livestock, and food and cash crops, working with farmers' groups rather than individual 'model' farmers, and with women just as readily as with men. At district level, there will be subject-matter specialists that the extension agent can call on. The agent will serve as a two-way conduit passing down messages from above, and passing up information from farmers on local problems and adaptations. The farmers themselves will be the last line of adaptation, combining new approaches and inputs with old methods, in the right places and proportions to suit their own circumstances.

Africa's labour shortage will have to be tackled if food production is to increase. As we have seen, this can seriously depress production by delaying planting, weeding or harvesting. It can create problems for conservation works which demand a lot of labour. Mechanization has classically been seen as the answer. In the long term it may well be, though it will have to be carefully designed to suit the special needs of Africa's environment. So far, mechanization has raised more problems than it has solved. Not only has it damaged the environment in many areas, but by focusing on one operation – usually ploughing – it has actually increased labour bottle-necks at other stages.

Over the next decade or two, improved hand tools and wider use of animal power are likely to give better results. The spread of trypano-tolerant livestock in humid areas will ease the labour shortage there. The most cost-effective first step in improving labour efficiency is the formation of farmers' groups. As we saw in Zimbabwe (p. 92), these can pool tools and draught animals and save wasteful duplication on tasks like livestock herding, food preparation, or going to market. Low-cost improvements in health, meanwhile, can increase workers' productivity.

EASING WOMEN'S BURDENS

One of the most pressing tasks of all is to deal with the many problems of women, estimated to be responsible for 70 per cent of subsistence food production in Africa. They need equal rights with men in landholding and legal matters, so that they have an incentive to help with conservation work on their husbands' land. They need the power to enter into contracts for credit or supplies, and to help to decide on the crops that are grown. They need equal attention from extension workers, almost always men who choose other men as model farmers, and special provisions in credit programmes.

Most immediately, they need help to reduce the crushing burden of work, imposed by the sexual division of labour in Africa and the migration of so many men to work in mines, plantations and cities. In the long run changes in men's attitudes are needed, and government example and exhortation will help. But customs change slowly, and faster-acting measures are urgently needed. Farmers' groups can ease the situation in farming. In the home, three major tasks together consume an average of around three hours a day. The task of firewood collection can be eased by tree-planting and fuel-efficient stoves, fetching water by more wells closer to home, and donkeys or carts to carry the cans. One of the most time-consuming tasks of all is pounding or grinding grain or roots, which may take an hour and a half of strenuous labour or an equally long trek to the mill and back. Cheap iron grinders – one or two per village at first – can make an enormous difference.

Male migration to work is directly linked to poverty – the smaller the farm, the more likely is the man to be away working. Improved yields and improved prices for agricultural produce will allow more men to stay at home, while the creation of jobs in rural areas – in low-cost road building, or small industries – would enable them to find paid work closer to home.

All these measures can be viewed as the first stage of an African Green Revolution. This revolution would, in its early years, be ultra-low-cost, and would involve virtually no imports. It would slow soil erosion at the same time as it boosted food production. It would be, to begin with, an *organic* agricultural revolution. Agriculturalists often point out that there are no free lunches, and that you can get nothing extra out of the soil without putting more in. *But there are a number of potential free lunches that can improve soil fertility without any purchased inputs*. Biological nitrogen fixation by legumes is one of these. Trees, whose deep roots tap reserves of water and nutrients inaccessible to annual crops, are another. Organic residues, so often burned on site or as fuel, are a third. (See chart, pp. 340–1.)

This first stage is not one that can be skipped, even by the few wealthier countries that can afford to import fertilizer or agricultural machinery for the foreseeable future. *Soil and water conservation, organic inputs, and trees, are not the poor man's substitute for chemicals and machinery. Given Africa's climate and soils, they are the indispensable foundation and pre-condition of sustainable agriculture.*

On the other hand, these early steps alone will not be enough to feed Africa's rapidly expanding populations. They will buy a breathing space, perhaps of a decade or so. Chemical fertilizers and improved seeds and breeds will gradually be brought in; hardy seeds bred for their ability to perform in farmers' circumstances, at first open-pollinated varieties rather than hybrids which have to be distributed every year. Chemical fertilizer can start in most areas with modest doses of phosphates which have the most dramatic impact on yields, followed later by carefully chosen nitrogenous and potash fertilizers.

The early stages will prepare the ground for later stages. Soil and water conservation and organic inputs increase the impact of chemicals and improved seeds. As farmers gain confidence in extension workers' advice, and cash income from increased production, they will be more willing and able to invest in inputs that provide less spectacular returns. The country's extension system can then be supplemented with a network, preferably co-operative or private, to supply purchased inputs. The savings on food imports will allow more imports of agricultural inputs.

REDUCING POPULATION PRESSURE

No one should underestimate the size of the task ahead. If Africa is to progress at all, agricultural production must grow faster than population. Simply to keep neck and neck, production would have to grow at 3.2 per cent a year until the end of the century. To provide improved incomes for the farming majority, agricultural production would need to expand by at least 4 per cent a year. That is twice as fast as the rate Africa achieved during the 1970s.

The low-cost measures we have outlined, phasing gradually into higher-input agriculture, will go a long way. Anything that improves the sustainable yield of the land helps to slow down the increase in population pressure. With higher yields, less new land would need to be cleared for cultivation, more could be left as forest and pasture. But the slow-down in pressure would clearly be more dramatic if Africa could reduce her rate of population growth. The smaller the future populations, the lower will be the requirement for cropland, the lower the demand for fuelwood, and the lower the animal population. With fewer heirs to divide the land between, farm sizes will not shrink so fast, so the chance of obtaining a decent family income from the average holding will be better.

Reduced population growth can dramatically cut the scale of effort required in agriculture. As we have seen, by the year 2025 Africa will be able to feed only 40 per cent of her expected population of 1640 million (the United Nations *medium* projection) if she is still using low inputs. This means she would have to import 60 per cent of her food requirements – or alternatively, increase food production by 150 per cent above the low input level. If the birth rate was, say, an average of 9 points lower over the period, Africa's 2025 population would be only 1140 million – a whole 1980-size Africa less than the medium projection. In this case it could feed 58 per cent of its population using low inputs. It would need to import enough food for 480 million people – less than half as many as with the higher population total – or to increase food production by only 72 per cent above the low input level. *A reduction of only 20 per cent in the birth rate would cut the effort required in food production or purchase by more than 50 per cent.*

Achieving such a drop, however, is easier said than done. As we have seen in Chapter 14, the direct approach to lowering the birth rate in Africa, through availability of modern contraceptives, is not very promising. The cost of making these available to scattered rural populations is high, the import requirements high, the managerial requirements high. Even if they were universally available, most African couples want large families and would probably use contraceptives to replace abstinence or withdrawal, rather than to reduce family size.

The approach, for most countries, will have to be primarily indirect, aimed at influencing the factors that affect the size of family people want: reducing women's workload so that they don't need extra hands to help, increasing enrolment in primary and secondary education, which reduces children's availability for labour, and increases their costs in uniforms or school fees; giving women rights to a portion of their husband's land which they would keep on divorce or widowhood; reducing vulnerability to drought through water conservation and improved food security; and finally, and most importantly, through a reduction in infant and child mortality.

The most effective strategy here exactly parallels the approach in agriculture. First create a network of village health agents and trained traditional midwives, with village committees and community groups to back them. Then use these to spread ultra-low-cost techniques focused on self-help, requiring a minimum of outside or imported inputs: teaching people to monitor their children's growth using growth charts, to make oral rehydration salts from salt, sugar and boiled water, and water-filters from charcoal, sand and gravel, and to dig household latrines.

A reduction in infant mortality, of course, carries the danger of a temporary rise in the population growth rate, because it takes people longer to change the complex of attitudes and behaviours that affect the birth rate. However, the sooner the reduction in infant mortality comes about, the earlier the birth rate will fall, and the smaller will be the eventual stable population.

To minimize the temporary rise in population growth, heavy emphasis should be placed on those health actions that reduce fertility and improve health at the same time: discouraging pregnancies among women under twenty and over thirty, encouraging breast-feeding for at least twenty months and the spacing of children by at least three or four years.

Once again, these interventions could be regarded as the first stage of a family welfare revolution providing the indispensable basis of popular awareness and self-help on which health care can flourish. Other measures, from vitamin A pills, immunization and use of modern medicines and contraceptives, to cement-lined wells and water pumps, can be added gradually as national resources and aid permit.

LEARNING TO LIVE DANGEROUSLY

Two powerful factors in Africa's future lie largely outside her own control: the climate, and the international economy.

The threat of drought hangs a sword over all scenarios for the dozen or so countries of the Sahel. The last two years have seen a break in the long dry spell that began around 1968. Whether they herald the end of drought, or merely an interlude in a longer decline, no one can say. 'Our current long-term forecasting capability is close to zero,' British climatologists G. Farmer and T. Wigley admit.

During the 1973–74 drought, most Sahelian governments hoped and prayed that good rains would return. After the 1983–84 drought, the mood has changed. Most experts in and out of Africa now agree that it is safer to plan on the assumption that years of low rainfall will be the norm, than to count on them coming to an end.

There is little or nothing that Africans can do to reduce the incidence of drought. If the cause is some long-term shift in the global climate caused by variations in the earth's orbit or the sun's output, then the Sahara will advance, as it has done in the distant past, and human beings will sooner or later be forced to retreat. If the cause is the warming 'greenhouse' effect of raised carbon dioxide levels in the atmosphere due to burning of fossil fuels, there is not much Africa alone could do to reduce emissions of the gas. In any case the connection is far from clear: previous episodes of global warming have been associated with a *wetter* Sahel.

If drought cannot be controlled, then the only possible approach is to adapt to it: above all to reduce farmers' *vulnerability* to drought, so that they can withstand a year or two of low rainfall without the risk of famine. This will mean small-scale irrigation; measures that increase infiltration and improve the water-holding capacity of the soil; encouragement of intercropping, planting of fodder, fruit and nut trees, whose deeper roots can tap lower levels of groundwater; and fostering deep rooting in crops by breeding and by use of phosphate fertilizers. Plants can be bred for early maturity, so that they produce even in a short rainy season. All projects and innovations outside the humid zone should be subjected to a risk-sensitivity analysis, to find out what the results would be at lower levels of rainfall, and what the risks of lower rainfall are. As a general rule, *no innovation or package should be offered that would leave farmers worse off in a bad year than they would be with their traditional techniques and varieties.*

Even with all these measures, there will inevitably be years or runs of years where drought does hit home, and the only possible answer is to stand ready to soften the blow. Disaster response lies outside the scope of this book. But it is clear that the system should not depend on harrowing television images of famine, generating charity donations and political pressures on donor governments, before it moves into action. By the time such pictures are there for the shooting, it is already far too late for sensible steps. What is needed instead is a virtually automatic system of warnings and graduated responses rising from national to regional and international level. Whenever possible, cash or food aid should be given in the home village, in exchange for work on schemes designed to reduce vulnerability to drought and famine.

In the long run, speedy advances in Africa will require changes in the world economic order. She will need some stabilization of the wild fluctuations of commodity prices, with their disruptive influence on the balance of trade and government budgets. For non-perishable items, buffer stocks, backed by funds that can buy and sell supplies, can even out some of the fluctuations in supply and demand for less perishable commodities. Commodity pacts like OPEC can, if they achieve agreement, regulate the supply. Compensatory grants and loans can make up for short-falls in commodity earnings and cushion the blows of the commodity market.

Africa would also benefit from measures that might increase the price of her commodity exports in relation to import prices. These would involve far-reaching reforms of the agricultural systems of Western countries, especially the Common Market, with its artificially high internal prices, tariffs and quotas against competing imports, and massive surplus stocks and dumping operations. All this keeps down the world price of commodities that are produced in the North and the South, including cereals, vegetable oils, and sugar. Opening frontiers to freer trade in agriculture would benefit not only producers in the South, but consumers in the North as well.

African countries' earnings from commodities, and their pros-pects for industrialization, would be improved if they were able to export less in the raw state and do more of the processing themselves. At present, Western tariffs grow progressively steeper the more commodities are processed. Those tariffs should be gradually lowered, especially on products of interest to Africa.

Again, Western consumers would benefit as well as African farmers and factory workers.

There are, unfortunately, massive practical and political obstacles to these changes. They must be fought for and worked towards, but Africa cannot rely on them to help her out of her present predicament.

There are slightly better chances of an increase in aid or debt relief, though neither is any substitute for reforms in the international economic system that would give Africa the opportunity to earn and pay her own way. The debt crisis has put an unprecedented strain on African economies. The oil price drop has eased the situation a little, but increased aid will still be needed to reduce debt's dampening effect on growth.

International reforms will help, but Africa need not and must not wait on these. In spite of all the current budgetary and balance of payments problems, there is much that can still be done using low-cost, low-import, self-help approaches, supplemented by judicious aid.

A dozen years ago, most people were convinced that India faced a dark Malthusian future of catastrophe. Today India is self-sufficient in food. Africa too can surprise the world. It will be an uphill fight, but it can be won. The evidence of some of the most successful projects in Africa shows that it is possible. *It is a future that we know can work*. Africa can pull herself through. Her farmers have all the skill and adaptability and energy required. Her leaders and the international community must supply the framework in which they can fulfil their potential.

A GREEN REVOLUTION FOR AFRICA

Stage one: The Foundations

1. Improvement of producer prices for food and cash crops plus devaluation or currency de-regulation
2. Security of tenure and use of land, trees, pasture including full title for women
3. Functional integration of agriculture, livestock, forestry and conservation

- Farming systems focus of research
- Rapid creation of network of multi-purpose agricultural extension agents using

- pyramid training
- Formation of farmers' groups
- Focus on food crops and women
- Production emphasis in conservation, and vice-versa

4. Dissemination of no-cash cost and minimal-cost techniques with high returns to labour, capital, land

- *Conservation of soil, water and nutrients*
  Stone lines
  Trash lines
  Planted bunds
  Grassed strips
  Contour ridges
  Infiltration zones
  Terracing (only on steepest slopes)

- *Plus, for semi-arid areas*
  Water-harvesting
  Water-spreading
  Windbreaks
  Dune fixation

- *Improvement of soil fertility and structure*
  Nitrogen fixation by legumes in intercrops and rotations
  Mulching with crop residues
  Planted fallow of leguminous fodder or mulch crops

- *Forestry*
  Village management of natural forest.
  Agroforestry with local varieties: multi-storey farming (humid)

  alley cropping
  acacias (semi-arid)
  tree crops
  living fences
  close hedges for terrace formation emphasis on multi-purpose trees:
    fruit and nuts, fodder, mulch, nitrogen fixation, timber, poles, stakes, fibres, fuelwood

- *Small scale irrigation*
  Valley bottoms
  Flood plains
  Earth dams

- *Improved timing and spacing of planting*

- *Livestock measures*
  Fodder banks of legumes and browse trees
  Increased use of camels and goats (semi-arid)
  Alley cropping for goats and sheep (humid)
  Hand-dug wells, dispersed water points
  Controlled grazing in farming areas

- *Improved labour availability and productivity*
  Through reduced women's burdens:
  Fuelwood from agroforestry
  Improved 3-stone stoves

Closer water points
Farmers' groups to pool oxen and tools
Improved health
Child-spacing

Stage two:

Low-cost techniques that require some nationwide distribution service, minimal imports and small cash investments

- Improved crop varieties (open-pollinated) bred for performance under peasant conditions
- Phosphate fertilizers as booster for legumes
- Introduced tree species (*Leucaena* etc)
- Cross-breeding of livestock
- Low-cost tsetse traps
- Multiplication of trypano-tolerant livestock

- Cheap wells
- Hand-operated pumps
- Scooped ponds
- *Improved devices to reduce women's burdens*
  Hand grinders
  Shellers
  Plus, for semi-arid areas
  Ploughing – one-ox, chisel
  Tied ridging

Stage three:

Moderate-cost techniques involving some imports, higher cash investment by farmers, and competent nationwide supply and maintenance systems

- Hybrid seeds
- Moderate levels of adapted nitrogenous and potash fertilizers
- Selective mechanization of labour bottle-necks
- Washbores

- Motorized pumps
- Biogas using bargain technology, e.g. polythene sausage
  Plus, for humid and sub-humid areas
  Minimum tillage with herbicides

NB:
1. Each stage builds on and incorporates the previous stages, which create the surplus and capital needed for the later stages.
2. Above list is suggestive, and not exhaustive. All techniques and packages must be tried and adapted nationally and locally with strong farmer participation.

Maps

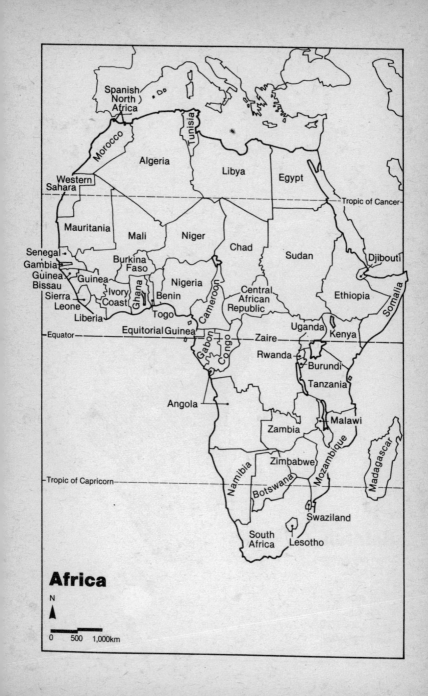

Spanish
North
Africa

Morocco

Tunisia

Algeria

Libya

Egypt

Western
Sahara

Tropic of Cancer

Mauritania

Mali

Niger

Chad

Sudan

Djibouti

Senegal

Gambia

Guinea
Bissau

Guinea

Burkina
Faso

Nigeria

Ethiopia

Somalia

Sierra
Leone

Ivory
Coast

Ghana

Benin

Cameroon

Central
African
Republic

Liberia

Togo

Equitorial Guinea

Uganda

Kenya

Equator

Gabon

Congo

Zaire

Rwanda

Burundi

Tanzania

Angola

Zambia

Malawi

Mozambique

Madagascar

Namibia

Zimbabwe

Tropic of Capricorn

Botswana

Swaziland

South
Africa

Lesotho

Africa

N

0   500   1,000km

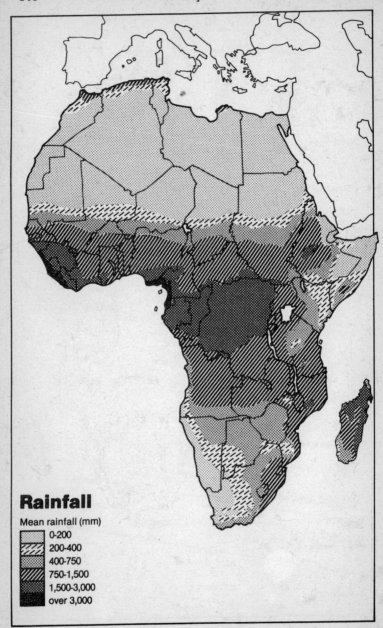

Rainfall

Mean rainfall (mm)

0-200
200-400
400-750
750-1,500
1,500-3,000
over 3,000

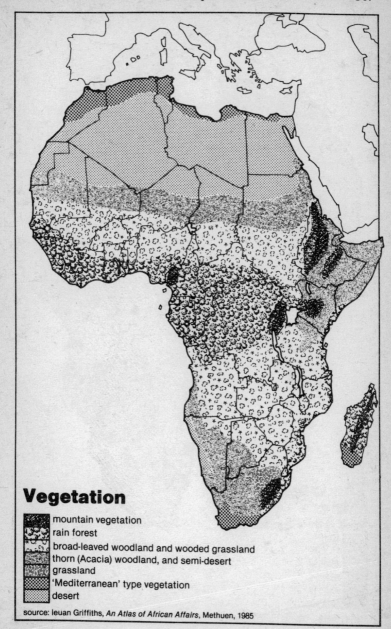

Vegetation

- mountain vegetation
- rain forest
- broad-leaved woodland and wooded grassland
- thorn (Acacia) woodland, and semi-desert
- grassland
- 'Mediterranean' type vegetation
- desert

source: Ieuan Griffiths, *An Atlas of African Affairs*, Methuen, 1985

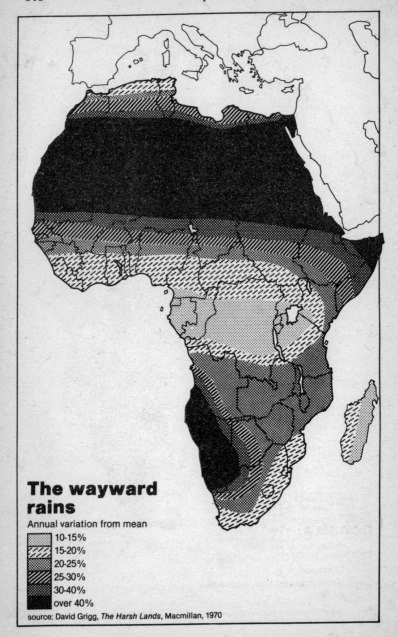

The wayward rains

Annual variation from mean

- 10-15%
- 15-20%
- 20-25%
- 25-30%
- 30-40%
- over 40%

source: David Grigg, *The Harsh Lands*, Macmillan, 1970

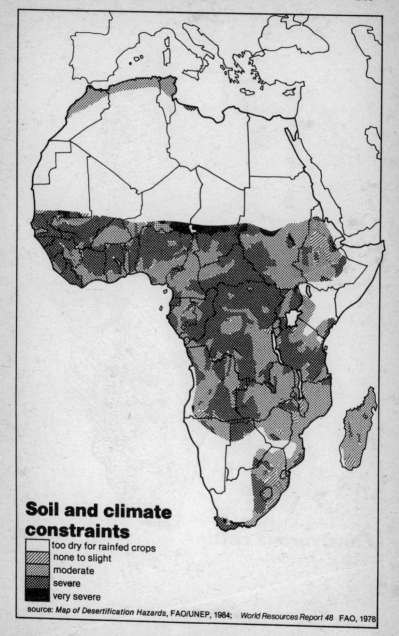

Soil and climate constraints

	too dry for rainfed crops
	none to slight
	moderate
	severe
	very severe

source: *Map of Desertification Hazards*, FAO/UNEP, 1984; *World Resources Report 48* FAO, 1978

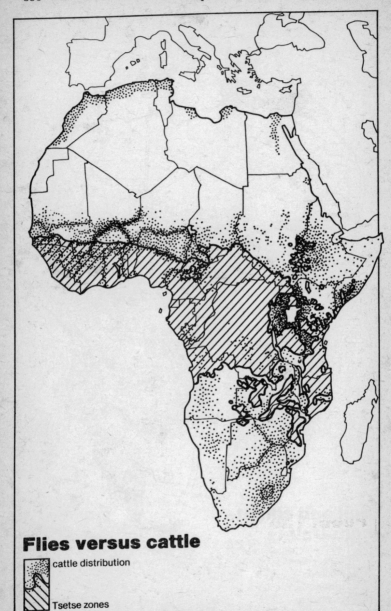

Flies versus cattle

cattle distribution

Tsetse zones

source: *Crisis of Sustainability*, AGD 801/1/, FAO, 1985

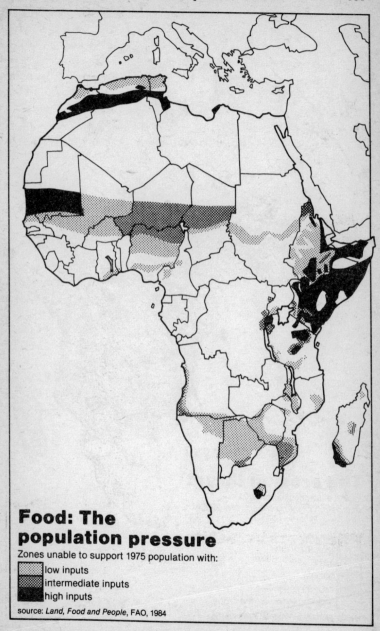

Food: The
population pressure

Zones unable to support 1975 population with:

low inputs
intermediate inputs
high inputs

source: *Land, Food and People*, FAO, 1984

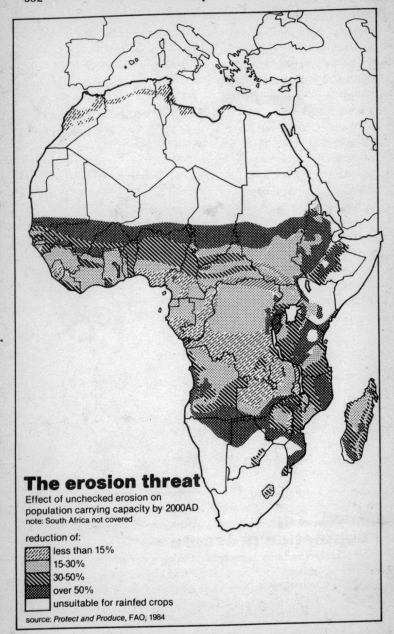

The erosion threat

Effect of unchecked erosion on
population carrying capacity by 2000AD
note: South Africa not covered

reduction of:

- less than 15%
- 15-30%
- 30-50%
- over 50%
- unsuitable for rainfed crops

source: *Protect and Produce*, FAO, 1984

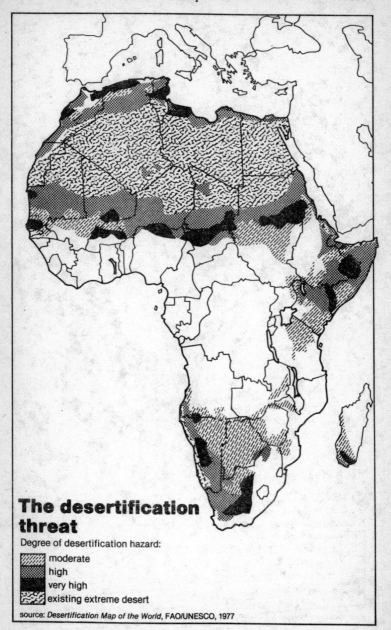

The desertification threat

Degree of desertification hazard:

- moderate
- high
- very high
- existing extreme desert

source: *Desertification Map of the World*, FAO/UNESCO, 1977

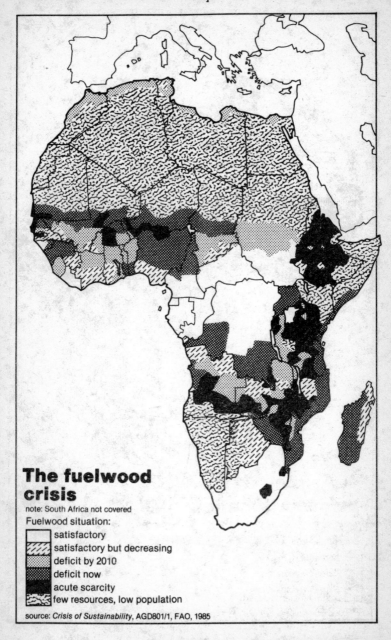

The fuelwood crisis

note: South Africa not covered

Fuelwood situation:

- satisfactory
- satisfactory but decreasing
- deficit by 2010
- deficit now
- acute scarcity
- few resources, low population

source: *Crisis of Sustainability*, AGD801/1, FAO, 1985

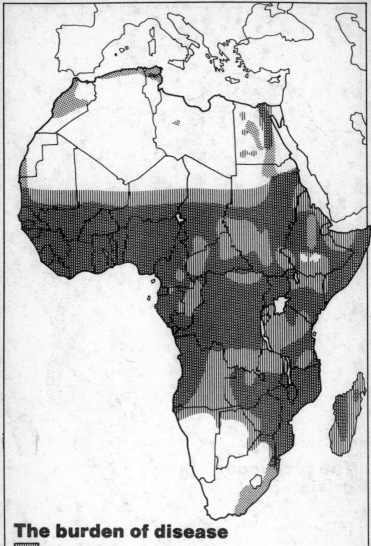

The burden of disease

 areas with malaria

areas with both

areas with bilharzia

source: Ieuan Griffiths, *An Atlas of African Affairs*, Methuen, 1985

Bibliography and References

1 THE DIMENSIONS OF CRISIS

Drought impact: this section is based on *Africa Emergency Report* nos. 1–5, UN Office of Emergency Operations for Africa, United Nations, New York, 1985; background papers for the Geneva Conference on the emergency in Africa, March 1985; FAO, *Food Situation in African Countries Affected by Emergencies*, special monthly reports, January 1985 – April 1986; and UNICEF, *Within Human Research*, UNICEF, New York, 1985. For an overview, see Timberlake, Lloyd, *Africa in Crisis*, Earthscan, London, 1985.

For population references, see Chapter 14.

Food production: trends in food production are calculated from *FAO, Production Yearbooks, 1981* and *1984*, FAO; *The State of Food and Agriculture* 1982 and 1984 (SOFA), and FAO, *1985 Country Tables*. Country trends from SOFA 1982 and 1984.

Cereal production, imports and self-sufficiency ratios calculated from FAO, *Food Outlook 1982* and *1984 Statistical Supplements* and *Food*, February 1986 issue (for 1984 and 1985).

Other data from World Bank, *Financing Adjustment with Growth in Sub-Saharan Africa 1986 – 90*, World Bank, Washington DC, 1986.

Debt figures are from World Bank, *World Debt Tables*, first supplement, 1984 – 5, World Bank, Washington DC, 1985, and from World Bank, *World Development Report 1985* and *1986*.

Commodities: commodity export dependence: World Bank, *Accelerated Development in Sub-Saharan Africa*, World Bank, 1981. Commodity price trends from UN Conference on Trade and Development, *UNCTAD Statistical Pocketbook*, UNCTAD, New York, 1984. Terms of trade and export volume from *WDR 1985*, op. cit.

Industrial strategy: see World Bank, *Accelerated Development*, op. cit. Figures for growth of production and shares in GDP from *WDR 1985*, op. cit.

Poverty: incidence from Hopkins, Malcolm, 'Employment Trends in Developing Countries', *International Labour Review*, Vol. 122 no. 4. Growth of per capita incomes from *World Development Report 1985*. Malnutrition incidence from FAO, *Fifth World Food Survey*, FAO,

Rome, 1985. Calorie intakes from FAO, *State of Food and Agriculture 1982* and *1984*, Rome, 1983 and 1985.

Rural poverty income differentials derived from Hopkins, Malcolm, op. cit.; rural-urban poverty levels from Bequele, Assefa and Van der Hoeven, Rolf, 'Poverty and Inequality in Sub-Saharan Africa', *International Labour Review*, vol. 119 no 3.

Agrarian Situation: landlessness from Sinha, Radha, *Landlessness*, FAO, Rome, 1985. Distribution of holdings from Parthasarathy, G, *Understanding Agriculture* (mimeo), FAO, 1979.

2 THE HARSHEST HABITAT

Climate: for trypanosomiasis, see Chapter 13. Nicholson, Sharon, *The Sahel: A Climatic Perspective*, OECD (CILSS), Paris, 1982; Farmer, G., and Wigley, T., *Climatic Trends for Tropical Africa*, School of Environmental Science, University of East Anglia, Norwich, 1985; FAO, *Crisis of Sustainability: Africa's Land Resource Base*, AGD/801/1, FAO, Rome, November 1985; Harrison, Paul, *Land, Food and People*, FAO, Rome, 1984; Kandel, Robert S., *Mechanisms Governing the Climate of the Sahel*, Club du Sahel, OECD, 1984; Brown, Lester and Wolf, Edward C., *Reversing Africa's Decline*, Worldwatch Paper 65, Worldwatch Institute, Washington DC, 1985.

Soils: UNEP, *Map of Desertification Hazards*, explanatory note, Nairobi, 1984; *FAO-UNESCO Soil Map of the World*, vol. 6, UNESCO, Paris, 1977; FAO, *Crisis of Sustainability*, op. cit.; Higgins, G.M. *et al*, *Potential Population Supporting Capacities of Lands in the Developing World*, FAO, Rome, 1982; erodibility and erosivity: Hudson, Norman, *Soil Conservation*, Batsford, London, 2nd edition, revised, 1985; crusting: Barber, R., *An Assessment of the Dominant Soil Degradation processes in the Ethiopian Highlands*, Ethiopian Highlands Reclamation Study, Working Paper 23, Addis Ababa, 1984; Young, Anthony, *Tropical Soils and Soil Survey*, Cambridge University Press, Cambridge, 1976.

Shifting cultivation: FAO, *Shifting Cultivation and Soil Conservation in Africa*, Soils bulletin no. 24, FAO, Rome, 1974; FAO, *Changes in Shifting Cultivation in Africa*, Forestry Paper no. 50, FAO, Rome, 1984.

3 WHY THINGS GO WRONG IN AFRICA

This chapter is based on a number of general reviews of the dimensions and causes of project failure. In particular, World Bank, *Tenth Annual Review of Project Performance Audit Results 1984*, Washington DC, 1985; Berg, Robert, *Foreign Aid in Africa*, Committee on African Development Strategies, Washington DC, 1984, plus a number of more detailed sectoral surveys, especially World Bank, *Desertification in the Sahelian and Sudanian Zones of West Africa*, Report no. 5210, Western Africa Projects

Department; Sandford, Stephen, *Review of World Bank Livestock Activities in Dry Tropical Africa*, World Bank, Washington DC, 1981; Weber, Fred, *Review of CILSS Forestry Sector*, Forestry Support Programme, USAID, Washington DC, 1982; USAID Africa Bureau, *Report of Workshop on Forestry Programme Evaluation*, USAID, Washington DC, 1984; and Steinberg, David, *Irrigation and AID's experience*, Programme Evaluation Report no. 8, USAID, Washington DC, 1983.

Examples of the proliferation of aid projects are from World Bank, *World Development Report 1985*, World Bank, Washington DC, 1985, p. 107.

African colonial history see Crowder, Michael, *West Africa under Colonial Rule*, Hutchinson, London, 1968; Kjekshus, Helge, *Ecology Control and Economic Development in East African History*, University of California Press, Berkeley, 1977; Palmer, Robin and Parsons, Neil, *The Roots of Rural Poverty in Central and Southern Africa*, Heinemann, London, 1977. For a general history, Curtin, Philip, *et al*, *African History*, Little, Brown and Company, Boston, 1978. On modern conflicts and military spending, Sivard, Ruth Leger, *World Military and Social Expenditures 1985*, World Priorities, Washington DC, 1985.

Urban bias and price policy: see Lipton, Michael, *Why Poor People Stay Poor*, Temple Smith, London, 1977; income differentials calculated from Hopkins, Malcolm, 'Employment Trends in Developing Countries', *International Labour Review*, vol. 122 no. 4; urban population trends from UN Population Division, *World Population Prospects, Estimates and Projections as assessed in 1982*, United Nations, New York, 1985; on exchange rate and trade policy, see World Bank, *Financing Adjustment with Growth*, World Bank, Washington DC, 1986; World Bank, *Towards Sustained Development in Sub-Saharan Africa*, Washington DC, 1984; and World Bank, *World Development Report 1986*, Washington DC, 1986.

Social Constraints: World Bank, *Accelerated Development in Sub-Saharan Africa*, World Bank, Washington DC, 1981; Parthasarathy, G., *Understanding Agriculture* (mimeo), FAO, March 1979; Ruthenberg, Hans, *Farming Systems in the Tropics*, Clarendon Press, Oxford, 1980; Lele, Uma, *The Design of Rural Development*, Johns Hopkins, Baltimore, 1975.

Women: Economic Commission for Africa, *The Role of Women in African Development*, E/CONF 66/BP/A, ECA, Addis Ababa; Mitchnik, David, *The Role of Women in Rural Zaire and Upper Volta*, Oxfam, 1977; Boserup, Ester, *Woman's Role in Economic Development*, Allen and Unwin, London, 1970; Economic Commission for Africa, *Role of Women in African Development*, International Women's Year Conference, E/conf/66/8, Mexico, 1975.

4 TAPPING THE POTENTIAL

On traditional farming systems see Richards, Paul, *Indigenous Agricultural Revolution*, Hutchinson, London, 1985; Kjekshus, Helge, *Ecology Control and Economic Development in East African History*, University of California Press, Berkeley, 1977; Lagemann, Johannes, *Traditional Farming Systems in Eastern Nigeria*, Weltforum Verlag, Munich, 1977; Ruthenberg, Hans, *Farming Systems in the Tropics*, 2nd Edition, Clarendon Press, Oxford, 1980; Allan, W, *The African Husbandman*, Oliver and Boyd, Edinburgh, 1965. For the Konso, Goettsch, Eggert *et al*, 'Crop Diversity in Konso Agriculture', *PGRC/ILCA Newsletter*, International Livestock Center for Africa, Addis Ababa, 1985. For the Dogon, see *Assignment Children no 45/ 6*, UNICEF, Geneva, 1979. The Chagga system is summarized in Goldsmith, E., and Hildyard, N., *The Social and Environmental Effects of Large Dams*, Wadebridge Ecological Centre, 1984.

5   BOOSTING FOOD PRODUCTION

Regional production trends are calculated from country tables in FAO, *State of Food and Agriculture 1982* and *1984*; the figures given are averages of country results, not weighted for populations.

Data on cash crop and food crop acreages and yields are calculated from FAO, *Production Yearbook 1981* and *1984*, FAO, Rome, using the Africa Developing Market Economies category.

Land reserves are from FAO Economic and Social Policy Department, *Agriculture: Towards 2000 Statistical Compendium*, FAO, Rome, 1983.

Regional fertilizer use is from FAO, *Country Tables 1985*, and country figures from FAO, *Fertilizer Yearbook vol. 31*, FAO, Rome, 1982. Cash crop dominance in fertilizer, data from FAO, *Atlas of African Agriculture*, FAO, Rome, 1986. The population density factor in agricultural change is discussed in Boserup, Esther, *The Conditions of Agricultural Growth*, Allen and Unwin, London, 1965.

Government spending on agriculture is from FAO, *State of Food and Agriculture 1983* (p.43); administration's share from *Development Forum*, September 1985. More data in FAO, *Atlas of African Agriculture*, op. cit.

6   A GREEN REVOLUTION FOR AFRICA

This chapter is based substantially on personal interviews and visits to the International Institute of Tropical Agriculture, in Ibadan, the International Crop Research Institute for the Semi-Arid Tropics' Sahelian Centre, in Niger, IITA and ICRISAT stations in Burkina Faso, and on IITA's and ICRISAT's *Annual Reports* and *Research Highlights* since 1979.

Research handicaps: Matlon, Peter, *A Critical Review of Objectives, Methods and Progress in Sorghum and Millet Improvement*, Village Studies Report no. 15, ICRISAT, Burkina Faso, 1985, and World Bank, *Agricultural Research and Extension*, World Bank, Washington DC, 1985.

Research staff numbers: FAO, *Atlas of African Agriculture*, FAO, Rome, 1986.

Modest Beginnings: Kenyan maize: Johnson, Charles *et al*, *Kitale Maize*, Project Impact Evaluation no. 2, USAID, Washington DC, 1979. Sorghum: *ICRISAT Annual Report 1983*. Cowpeas: Singh, S.R., *Impact of IITA's Grain Legume Improvement Programme*, IITA, Ibadan, 1984 Cassava Akoroda, M.O., *et al*, *Impact of IITA Cassava Varieties in Oyo State*, IITA, 1985.

Fertility: risk of fertilization in dry areas: Matlon, Peter, *The Technical Potential for Increased Food Production in the West African Semi-Arid Tropics*, ICRISAT, Burkina Faso, 1983.

Phosphates: personal interviews at ICRISAT's Sahelian Centre.

Crop Residues: FAO, *Organic Materials as Fertilizers*, Soils Bulletin no. 27, FAO, Rome, 1975; FAO, *Organic Recycling in Africa*, Soils Bulletin no. 43, FAO, Rome, 1980; Lal, Rattan, *Role of Mulching Techniques in Tropical Soil and Water Management*, Technical Bulletin no. 1, IITA, Ibadan, 1975; impact on millet: personal communication from Michael Klaij, ICRISAT Sahelian Centre.

Nitrogen Fixation: Hamdi, Y.A., *Application of Nitrogen-fixing Systems in Soil Management*, Soils Bulletin no. 49, FAO, Rome, 1982.

Intercropping: Fussell, L.K., and Serafini, P.G., *Crop Associations in the Semi-Arid Tropics of West Africa*, unpublished draft, ICRISAT Sahelian Centre, Niamey, 1985. ICRISAT *Annual Reports 1979–1983*, Hyderabad, India.

7 GAINING GROUND: SUCCESS IN SOIL CONSERVATION

The extent of the problem: Brown, Lester and Wolf, Edward, *Soil Erosion*, Worldwatch Paper 60, Worldwatch Institute, Washington DC, 1984; Brown and Wolf, *Reversing Africa's Decline*, Worldwatch Paper 65, 1985; FAO, *Protect and Produce*, FAO, Rome, 1984; FAO, *Natural Resources and the Human Environment for Food and Agriculture in Africa*, FAO, Rome, 1982; FAO/UNEP, *A Provisional Methodology for Soil Degradation Assessment* (with maps), FAO, Rome, 1979; Harrison, Paul, *Land, Food and People*, FAO, Rome, 1984; FAO, *Report on the Second FAO/UNFPA Expert Consultation on Land Resources for Populations of the Future*, FAO, 1980; FAO, *Map of Desertification Hazards, Explanatory Note*, prepared for UNEP, Nairobi, 1984; Lal, Rattan, 'Soil Erosion' in FAO, *Shifting Cultivation and Soil Conservation in Africa*, Soils Bulletin no. 24, FAO, Rome, 1974; Blaikie, Piers, *The Political Economy of Soil Erosion in Developing Countries*, Longman, London, 1985; Reij, Chris, and Turner, Stephen, *Soil and Water Conservation in Sub-Saharan Africa*, International Fund for Agricultural Development, Rome, 1985.

Techniques of Conservation: Weber, Fred, and Hoskins, Marilyn, *Fiches Techniques de Conservation du Sol*, CILSS, 1983; FAO, *Organic Recycling in Africa*, Soils Bulletin no. 43, FAO, Rome, 1980; Unger, Paul, *Tillage Systems for Soil and Water Conservation*, Soils Bulletin no. 54, FAO, Rome, 1984; Lal, Rattan, *Role of Mulching Techniques in Tropical Soil and Water Management*, Technical Bulletin no. 1, International Institute of Tropical Agriculture, Ibadan, 1975; Lal, Rattan, *No-Till Farming*, Monograph no. 2, IITA, Ibadan, 1983; Hudson, Norman, *Soil Conservation*, Batsford, London, 2nd edition, revised, 1985.

Kenya: Permanent Presidential Commission on Soil Conservation and Afforestation, *Kenya's Efforts to Conserve Soil, Water and Forests*, Annually since 1982; Carl G. Wenner, *Soil Conservation in Kenya*, 7th Edition, Ministry of Agriculture, Nairobi, 1981; Ministry of Agriculture, *SIDA – Sponsored Soil Conservation Projects Annual Report 1983–4*, Nairobi, 1984; Hedfors, Lars, *Evaluation and Economic Appraisal of Soil Conservation in a Pilot Area*, Ministry of Agriculture, Nairobi, 1981; Holomberh, *Evaluation and Economic Appraisal of Soil Conservation in Kalia Sub-location*, SIDA, undated; Ministry of Agriculture, *Socio-Economic Aspects of Soil Conservation in Kenya*, Nairobi, 1985.

Ethiopia: The following restricted documents were drawn on for this section: *Summary Review of Project Ethiopia 2488*, World Food Programme, WFP/CFA 15/11, Add C8, March 1983; World Food Programme, *Report of Evaluation Mission 1985*; John, B.C., *Design and Construction of Soil Retention Bunds*, and *Construction of Stone and Stone-faced Earth Bunds*, Ministry of Agriculture, Addis Ababa, 1984; Yeraswork, Admassie and Gebre, Solomon, *Food for Work in Ethiopia, A Socio-Economic Survey*, Institute of Development Research, Addis Ababa, 1985; plus the following working papers of the Ethiopian Highlands Reclamation Study, all published by the Food and Agriculture Organization and the Ministry of Agriculture, Addis Ababa: Yeraswork, Admassie *et al*, *Report on the Sociological Survey* (December 1983); Archer, A.C., *Assessment of the Current Grassland and Forestland Situation* (March 1984); Tagoe, C., *A Tentative Review of Agriculture in the Highlands* (November 1983); Thomas, Donald B., *Soil and Water Conservation in the Ethiopian Highlands* (July 1984); Barber, R., *An Assessment of the Dominant Soil Degradation Processes in the Ethiopian Highlands*; Constable, M., *Summary Assessment*, (May 1985).

8 FIGHTING THE DESERT: LAND RECLAMATION

Desertification: UN Conference on Desertification, *World Map of Desertification*, Nairobi, 1977; Mabutt, Jack, 'A New Global Assessment of the Status and Trends of Desertification', *Environmental Conservation*, Vol. 11 no. 2, 1984; Dregne, Harold, 'Combating

Desertification', *ibid*; Berry, Leonard, *Assessment of Desertification in The Sudano-Sahelian Region*, UN Environment Programme, GC12, Background Paper 1, May 1984; Grainger, Alan, *Desertification*, Earthscan, 1982; UN Environment Programme, *Map of Desertification Hazards*, UNEP, Nairobi, 1984.

Sand dune fixation: Maiga, Amadou N'Tirgny, *Evaluation des Projets Care à Bouza*, CARE, Niamey, 1983.

Guesselbodi Forest: Sève, Juan, and Tabor, J.A., *Land Degradation and Simple Conservation Practices*, 1985; Boudouresque, E., and Chase, R., *Vegetative Regeneration Trials on Bare Soils in the Region of Niamey*, 1984; Heermans, John G., *The Guesselbodi Experiment*, 1985 – all documents of the Forestry and Land Use Planning Project, USAID, Niger; ICRISAT Sahelian Centre, *Annual Report 1983*, Naimey, 1983.

Progress: *General Assessment of Progress in the Implementation of the Plan of Action to Combat Desertification*, 1978–84, UNEP GC12/9, Nairobi, 1984; *Summary Report on the Regional Seminar on Desertification*, Nouakchott, Club du Sahel, Paris, 1985.

9 EVERY DROP COUNTS: WATER CONSERVATION

Irrigation in Africa: extent, expansion and potential of irrigation from FAO, *Consultation on Irrigation in Africa*, Lomé, April 1986, AGL:IA/86/Docs I – A and I – C; FAO, *Production Yearbooks 1979* and *1984*, FAO, Rome; FAO, *Prospects and Trends of Irrigation in Africa*, 13th Regional Conference for Africa, ARC 84/6, FAO, 1984; CILSS, *The Development of Irrigated Agriculture in the Sahel*, Club du Sahel, 1980; Underhill, H.W., *Small-Scale Irrigation in Africa*, FAO, Rome, 1984; Bottrall, A.F., *Comparative Study of the Management and Organization of Irrigation Projects*, World Bank Staff Working Paper no. 458, Washington DC, 1981; Steinberg, David, *Irrigation and Aid's Experience*, Programme Evaluation Report no. 8, USAID, Washington, 1983; Benedict, Peter, *et al*, *Sudan: The Rahad Irrigation Project*, Project Impact Evaluation Report no. 31, USAID, 1982; Goldsmith, Edward, and Hildyard, Nicholas, *The Social and Environmental Impact of Large Dams*, Wadebridge Ecological Centre, 1984.

Dry Season Gardening: Lutheran World Relief, Niger, *Annual Report 1984–85 Season*, Niamey, 1985; Thacher, Edith and Fitzgerald, Bill, *Evaluation of LWR Projects in Niger*, LWR, Niamey, 1985.

Valley bottom development in Kano: World Bank, *Kano Agricultural Development Project, Staff Appraisal Report*, no. 3089a – UN, World Bank, Washington DC, 1981; Tarcisius, M., *Knarda's Approach to Small-scale Irrigation*, Kano State Agricultural and Rural Development Authority, Kano, 1985.

Waterharvesting etc: Wright, Peter, *La Gestion des Eaux de Ruissellement*, Oxfam, Oxford, 1984; Matlon, Peter S., *The Technical Potential for*

Increased Food Production in the West African Semi-Arid Tropics, ICRISAT, Burkina Faso, 1983; ICRISAT/Upper Volta Co-operative Programme, *Annual Report 1983*, Ouagadougou, 1984; National Academy of Sciences, *More Water for Arid Lands*, NAS, Washington DC, 1974; Postel, Sandra, *Conserving Water*, Worldwatch Paper 67, Worldwatch Institute, Washington DC, 1985.

10 SEEING THE WOOD AND THE TREES: FORESTRY

Deforestation: *Tropical Forest Resources,* Forestry Paper no. 30, FAO, Rome, 1982; Tropical Forest Resources Assessment Project, *Forest Resources of Tropical Africa*, vols. I – II, FAO, Rome, 1981; Myers, Norman, *Conversion of Tropical Moist Forests*, National Academy of Science, Washington, 1980.

Forestry Failures: Weber, Fred, *Review of CILSS Forestry Sector Programme Analysis Papers*, USAID Forestry Support Programme, Washington, 1982; World Bank Forestry Department, *Review of World Bank Financed Forestry Activity*, Washington, 1983 and 1984; *Report of Workshop on Forestry Programme Evaluation*, USAID Bureau for Africa, Washington DC, 1984; World Bank, *Deforestation, Fuelwood Consumption and Forest Conservation in Africa*, World Bank, Washington, 1986.

Natural Forest Management: Jackson, J.K. *et al*, *Management of the Natural Forest in the Sahel Region*, Club du Sahel, Paris, 1983; World Resources Institute, *The World's Tropical Forests*, WRI, Washington, 1985.

Majjia Valley: Dennison, Steve, Delehanty, James, Hoskins, Marilyn, Thomson, James, *Majjia Evaluation Study*, interim reports and sociological reports, CARE, New York, 1984 and 1985; Bognetteau – Verlinden, Else, *Study on Impact of Windbreaks in Majjia Valley*, thesis, Agricultural University, Wageningen, Holland, 1980; Maiga, Amadou N'Tirgny, *Evaluation des Projets Care à Bouza 1975–82*, CARE, Niger, 1983.

Kenya: Van Gelder, Barry and Kerkhof, Paul, *The Agroforestry Survey in Kakamega District*, Kenya Woodfuel Development Project Working Paper no. 6, Nairobi, 1985; Chavangi, Noel, *Cultural Aspects of Firewood Procurement*, KWDP Working Paper no. 4, Nairobi, 1984; Mathu, Winston, *Directory of Governmental and Non-Governmental Organizations working on Agroforestry*, Kenya Renewable Energy Development Project, Nairobi, 1985; Maathai, Wangari, *The Green Belt Movement*, Nairobi, 1985; *Proceedings of Agro-Forestry Workshop for High Potential Areas in Kenya*, Kenya Energy Non-Governmental Organizations Association, Nairobi, 1983; Permanent Presidential Commission on Soil Conservation and Afforestation, *Kenya's efforts to Conserve Soil, Water and Forests*, Nairobi, 1982, 1983, 1984.

11 TURNING FARMS INTO FORESTS: THE POTENTIAL OF AGROFORESTRY

General: Advisory Committee on the Sahel, National Research Council, *Agroforestry in the West African Sahel*, National Academy Press, Washington DC, 1983; FAO, *Changes in Shifting Cultivation in Africa*, Forestry Paper 50, FAO, Rome, 1984; *Improved Production Systems as an Alternative to Shifting Cultivation*, Soils Bulletin 53, FAO, Rome, 1984; World Bank, *Review of World Bank Financed Forestry Activity*, op. cit.; Weber, Fred, op. cit.; National Academy of Sciences, *Leucaena*, Washington DC, 1980 and 1983; NAS, *Tropical Legumes*, Washington DC, 1979; NAS, *Firewood Crops*, vols. 1 & 2, Washington DC, 1980 and 1983; Foley, Gerald, and Barnard, Geoffrey, *Farm and Community Forestry*, International Institute for Environment and Development, London, 1984; FAO, *Forestry for Local Community Development*, Forestry Paper 7, FAO, 1978.

Alley Cropping: Kang, B.T., Wilson, G.F., Lawson, T.L., *Alley Cropping*, International Institute of Tropical Agriculture, Ibadan, 1984; Kang, B.T., *Nitrogen Management in Alley Cropping Systems*, paper for International Symposium on Nitrogen Management, IITA, Ibadan, October 1984; Sumberg, J.E., *et al*, *An Economic Analysis of Alley Farming with Small Ruminants*, International Livestock Centre for Africa, Addis Ababa, 1985; Attah-Krah, A.N., *et al*, *Leguminous Fodder Trees in the Farming System*, ILCA Humid Zone Programme, Ibadan, 1985; Ngambeki, D.S., and Wilson, G.F., *Economic and On-farm Evaluation of Alley Farming*, IITA, Ibadan, 1984; *Research Highlights 1983*, IITA, Ibadan, 1984.

Agro-forestry in Kenya: project documents of the Kenya Renewable Energy Development Project, EDI/Ministry of Energy, PO Box 62360, Nairobi.

12 OF STICKS AND STOVES: SOLVING THE FUELWOOD CRISIS

Fuelwood Situation: FAO, *Map of the Fuelwood Situation in Developing Countries*, FAO, Rome, 1981; FAO, *Fuelwood Supplies in Developing Countries*, Forestry Paper no. 42, FAO, Rome, 1983; Eckholm, Eric, Foley, Gerald, Barnard, Geoffrey, Timberlake, Lloyd, *Fuelwood: the energy crisis that won't go away*, Earthscan, IIED, London, 1984.

Stoves: Foley, Gerald, and Moss, Patricia, *Improved Cooking Stoves*, Earthscan, IIED, London, 1983; FAO, *Wood for Energy*, Forestry Topics Report no. 1, FAO, Rome, 1984; Foley, Gerald, Moss, Patricia, and Timberlake, Lloyd, *Stoves and Trees*, Earthscan, IIED, London, 1984.

Kenya improved charcoal stoves programme: Hosier, Richard, *Patterns of Domestic Energy Consumption in Rural Kenya*, Beijer Institute

Fuelwood Working Paper, Nairobi, 1982; Kinyanjui, Max, *The Kenya Charcoal Stoves Programme, Interim Report*, Energy Development International, Nairobi, Kenya, 1984; Hyman, Eric, *The Strategy of Decentralized Production and Distribution of Improved Charcoal Stoves in Kenya*, Appropriate Technology International, Washington DC, 1985; Hyman, Eric, *The Economics of Fuel-Efficient Household Charcoal Stoves in Kenya*, ATI, Washington, 1985.

13   PASTORALISTS AND PLOUGHMEN: LIVESTOCK

General problems: Jahnke, Hans E., *Livestock Production Systems and Livestock Development in Tropical Africa*, Kieler Wissenschaftsverlag Vauk, Kiel, 1982; FAO, *The State of Food and Agriculture 1982*, FAO, Rome, 1983; Sandford, Stephen, *Management of Pastoral Development in the Third World*, John Wiley, New York, 1983. Livestock numbers are from FAO, *Production Yearbooks*, 1979 and 1984, FAO, Rome; UN Conference on Desertification, *Niger Case Study*, A/CONF/74/14, Nairobi, 1977; Child, R. Dennis *et al*, *Arid and Semi-Arid Lands*, Winrock International, Morrilton, 1984; Advisory Committee on the Sahel, *Environmental Change in the West African Sahel*, National Academy Press, Washington DC, 1984.

Programme problems: *Review of World Bank Livestock Activities in Dry Tropical Africa* (unpublished), World Bank, Washington DC, 1981; World Bank Operations Evaluation Department, *The Smallholder Dimension of Livestock Development*, Washington DC, 1985 (unpublished draft); Horowitz, Michael, *The Sociology of Pastoralism and African Livestock Projects*, Discussion Paper no. 6, Office of Evaluation, USAID, Washington DC, 1979; Institute for Development Anthropology, *Workshop on Pastoralism and African Livestock Development*, USAID, 1980.

Pastoralism: Cossins, N., *The Productivity and Potential of Pastoral Systems*, ILCA bulletin no. 21, 1985, International Livestock Centre for Africa, Addis Ababa, 1985; Lusigi, W., 'Desertification and Nomadism', *Nature and Resources*, January–March 1984; Swift, Jeremy (ed.), *Pastoral Development in Central Niger*, Ministry of Rural Development, Niamey, Niger, 1984; White, Cynthia, *Herd Reconstitution*, Pastoral Network Paper 18d, Overseas Development Institute, London, 1984.

Fodder banks: Le Houérou, H.N., *Browse in Africa*, ILCA, Addis Ababa, 1980; ILCA, *Annual Report 1983*; *ILCA Bulletin*, no. 19, 1984.

Mixed Herds: Wilson, R.T., *Benefits to Range Resource Use from Multiple Species Stocking*, ILCA, Addis Ababa, 1985 (mimeo); Wilson, R.T. (ed.), *Small Ruminants in African Agriculture*. ILCA, Addis Ababa, 1985.

Restocking: Hogg, Richard, *Restocking Pastoralists in Kenya*, Pastoral Network Paper 19c, ODI, London, 1985; Toulmin, Camilla, *Livestock Losses and Post-drought Rehabilitation*, ILCA, Addis Ababa, 1985.

Tsetse: ILCA, *Trypanotolerant Livestock in West and Central Africa*, ILCA monograph no. 2, ILCA, Addis Ababa, 1979; Trail, J.C.M., *et al*, *Productivity of Boran Cattle*, ILCA, Addis Ababa, 1985; International Laboratory for Research on Animal Diseases, ILRAD Reports, 1983–5, ILRAD, Nairobi.

For alley cropping with goats, see references to Chapter 11.

Kenya: see FAO, *State of Food and Agriculture 1982*, Rome, 1983, and Jahnke, op. cit.

Ethiopia: International Livestock Centre for Africa, *One-Ox Ploughs*, ILCA Bulletin no. 18, April 1984; *ILCA Newsletter* vol. 5 no. 1, 1986; Astatke, Abiye, *et al*, *Building Ponds with Animal Power*, ILCA, Addis Ababa, 1986.

14 TACKLING RUNAWAY POPULATIONS

Population data and projections are taken from UN Population Division, *World Population Prospects, Estimates and Projections as Assessed in 1982*, UN Department of International Economic and Social Affairs, New York, 1985.

Overpopulation: population densities are from World Resources Institute and International Institute for Environment and Development, *World Resources Report 1986*, Basic Books, New York, 1986. Desert and arid area from Higgins, G., *et al*, *Potential Population Supporting Capacities of Lands in the Developing World*, FAO, Rome, 1982; food and population study, ibid, summarized and analysed in Harrison, Paul, *Land, Food and People*, FAO, Rome, 1984, updated in FAO, *Crisis of Sustainability*, GGD/801/1, 1985, and FAO, *Consultation on Irrigation in Africa*, AGL:IA/86/Doc I–D, Lomé, 1986.

Levels of contraceptive use, fertility and ideal family size in Africa are from *Population Reports*, Series M no. 8, *Fertility and Family Planning Surveys*, Population Information Programme, Baltimore, October 1985.

Causes of high fertility: Bongaarts, John, *et al*, 'The Proximate Determinants of Fertility in Sub-Saharan Africa,' *Population and Development Review* vol. 10 no. 3, 1984; Faruquee, Rashid, and Gulhati, Ravi, *Rapid Population Growth in Sub-Saharan Africa*, Staff Working Paper no. 559, World Bank, Washington DC, 1983; *Sub-Saharan Africa: Population Pressures on Development*, Population Reference Bureau, New York, 1985.

Government policies: World Bank, *World Development Report 1984*, Washington DC, 1984; UN Population Division, *Report of the Fifth Population Enquiry among Governments*, UN Department of International Economic and Social Affairs, ESA/P/WP/83, New York, 1983.

Zimbabwe: internal project documents, plus Central Statistical Office, *Main Demographic Features of the Population of Zimbabwe*, Harare, 1985;

World Bank, Zimbabwe, *Population, Health and Nutrition Sector Review*, Report no. 4214–ZIM, World Bank, Washington DC, 1983; *Zimbabwe Reproductive Health Survey 1984,* Westinghouse Public Applied Systems, Maryland, 1985.

15   THE HUMAN RESOURCE

Most of this chapter is based on UNICEF, *State of the World's Children 1984, 1985* and *1986*, UNICEF, New York, and UNICEF, *Within Human Reach*, New York, 1985. See also *Assignment Children*, nos. 45/46 (water and sanitation), 55/56 (breast-feeding), 59/60 (community participation), 61/62 (child survival and development revolution), 69/ 70 (immunization).

Data on doctor availability are from World Bank, *World Development Report 1985*, World Bank, Washington DC; on water and sanitation coverage from *World Health Statistics*, vol. 39 no. 1, 1986; for the OECD survey, Miller, Duncan, *Self-help and Popular Participation in Rural Water Systems*, OECD, Paris, 1979; for basic services and primary health care, Djukanovic, V., and Mach, E.P., *Alternative Approaches to Meeting Basic Health Needs*, WHO, Geneva, 1975; WHO/UNICEF, *Primary Health Care*, WHO, Geneva, 1978.

Burkina Faso's programme is summarized in: *Vaccination Commando Burkina Faso*, UNICEF, Onagadongon, 1985.

16   BY THEIR OWN BOOTSTRAPS: MOBILIZING FOR CHANGE

This chapter is based on field visits plus mimeoed documents of the Naam movement (PO Box 100, Ouahigouya, Burkina Faso), and the Bale RIBS project. Zimbabwe's Organization of Rural Associations for Progress is described and analysed in Chavunduka, Dexter, *et al*, *Khuluma Usenza*, ORAP, PO Box 877, Bulawayo, Zimbabwe.

17   THE SECRETS OF SUCCESS

Apart from the conclusions of my own research, this chapter draws on the conclusions of the general and sectoral reviews of project failure cited under the bibliography to Chapter 3.

18   THE PROSPECTS FOR A QUANTUM LEAP

Population projections are from UN Population Division, *World Population Prospects as Assessed in 1982*, UN Department of International Economic and Social Affairs, New York, 1985; UN Secretariat, *Long Range Global Population Projections*, UN Population Bulletin no. 14;

projections for national plateau populations from World Bank, *World Development Report 1986*. Food projections from FAO, *Agriculture: Toward 2000*, FAO, Rome, 1982. Fuelwood projections from FAO, *Fuelwood Supplies in the Developing Countries*, Forestry Paper 42, FAO, Rome, 1983. Land use projections from FAO, *Atlas of African Agriculture*, FAO, Rome, 1986. Carrying capacity projections from Higgins, G.M., *et al*, *Potential Population Supporting Capacities of Lands in the Developing World*, FAO, Rome, 1982; and Harrison, Paul, *Land, Food and People*, FAO, Rome, 1984.

Contact addresses

The descriptions and drawings in this book are intended as a general introduction, not as a technical manual. It is essential that all measures should be tested locally and on-farm before wide dissemination. Readers wishing for technical data on the various techniques referred to, or more detailed information about the projects, should write to the following addresses.

GENERAL ADDRESSES:

Food and Agriculture Organization
Via delle Terme di Caracallà
00100 Rome
Italy

International Institute for Environment and Development
3 Endsleigh St
London WC1H 0DD

US office:
1717 Massachussetts Avenue NW
Washington DC 20036
USA

UN Environment Programme
PO Box 30552
Nairobi
Kenya

UNICEF
866 UN Plaza
New York
NY 10017
USA

USAID
Department of State
Washington DC 20523
USA

World Bank
1818 H Street NW
Washington DC 20433
USA

World Food Programme
Via delle Terme di Caracalla
00100 Rome
Italy

INTERNATIONAL
RESEARCH CENTRES:

**International Institute for
Tropical Agriculture (IITA)**
Oyo Road
PMB 5320
Ibadan
Nigeria

**International Crops Research
Institute for the Semi-Arid
Tropics (ICRISAT)**
Sahelian Centre
BP 12404
Niamey
Niger

**International Council for
Research in Agroforestry**
PO Box 30677
Nairobi
Kenya

**International Livestock Centre
for Africa**
PO Box 5689
Addis Ababa
Ethiopia

SPECIFIC CONTACTS:

[Addresses as above unless quoted
here]

ORGANIC RECYCLING: Charles
Ofori, FAO

MULCHING, MINIMUM TILLAGE: Dr
Rattan Lal, IITA

PHOSPHATE FERTILIZER: Dr André
Bationo, ICRISAT

INTERCROPPING: Humid zone – Dr
Humphrey Ezuma, IITA
Semi-arid zone – Dr Les Fussell,
ICRISAT

FANYA JUU TERRACES, TRASH LINES:
Soil and Water Conservation
Division
Ministry of Agriculture and
Livestock Development
PO Box 30028
Nairobi
Kenya

STONE LINES AGAINST CRUSTING,
HOSEPIPE WATER LEVEL:
Oxfam
274 Banbury Rd
Oxford
United Kingdom

MANAGEMENT OF NATURAL FOREST:
John Heermans
Forestry and Land Use Planning
Project
USAID
United States Embassy
BP 11201
Niamey
Niger

ALLEY CROPPING:
Dr B. T. Kang, IITA ICRAF

KENYA AGROFORESTRY RESEARCH,
CERAMIC CHARCOAL AND WOOD
STOVES:
Dr Amare Getahun
Kenya Renewable Energy
Development Project
PO Box 62360
Nairobi
Kenya

OX-DRAWN SCOOPS, ONE-OX
  PLOUGHS, PLASTIC BIOGAS
  DIGESTERS:
Frank Anderson
Ethiopian Highlands Programme
ILCA

NAAM MOVEMENT:
Dr Bernard Lédéa Ouedraogo
Association Six 'S'
BP 100
Ouahigouya
Burkina Faso

Index